Real-Time Control of the Industrial Enterprise

Real-Time Control of the Industrial Enterprise

PETER MARTIN AND WALTER BOYES

MOMENTUM PRESS

MOMENTUM PRESS, LLC, NEW YORK

Real-Time Control of the Industrial Enterprise

First published by Momentum Press®, LLC
222 East 46th Street, New York, NY 10017
www.momentumpress.net

ISBN-13: 978-1-60650-359-1 (hard cover, case bound)
ISBN-10: 1-60650-359-6 (hard cover, case bound)
ISBN-13: 978-1-60650-361-4 (e-book)
ISBN-10: 1-60650-361-8 (e-book)

DOI: 10.5643/9781606503614

Cover Design by Jonathan Pennell
Interior design by Exeter Premedia Services Private Ltd.,
Chennai, India

10 9 8 7 6 5 4 3 2 1

Printed in the United States of America.

Peter Martin: To my beautiful granddaughters—Karly and Lexi

Walter Boyes: To my wife Joy, and to all the dedicated workers in manufacturing and automation who have made this world a much nicer and easier place to live in.

Contents

List of Figures

List of Photographs

Foreword by Jim Davis

Today, Smart Manufacturing is in good company with a lot of smart areas. One can readily find Smart Manufacturing, Smart Buildings, Smart Energy Grid, Smart Oil and Gas Grid, Smart Machines, Smart Weapons, and more. In 2006, when Smart Manufacturing had its origins it was difficult to find Smart terminology in use anywhere. Over a span of only eight years, an initial flicker of "smart" and the potential for radical change in the manufacturing industry has been fanned into a steady flame of increasing clarity and brightness that comes with critical mass, definition, and direction. The vocabulary for Smart Technologies is now infused in many discussions including the Internet of Things (IoT), which offers terminology for the device, data, and communications that underpin what is *Smart.*

So, what is *Smart Manufacturing*? The manufacturing industry has already seen remarkable productivity gains over the past 40 years as a result of control and automation. To the Smart Manufacturing Leadership Coalition (SMLC), *Smart* is the sweet spot of several converging forces that are changing a 100-year-old mass-production model. Expectations for sustainability, energy, environmental impact, safety, and political performance are pushing more enterprise, integrated, and global targets while economic performance is rapidly giving way to on-demand value in the products and the materials from which they are made. The expectations for new and increased enterprise manufacturing performance metrics are not only changing but the need for change is accelerating. Manufacturing can no longer afford to optimize vertically without also addressing the "horizontal" effects across the enterprise and value chain. In fact, the greatest opportunities for *Smart* exist at the "seams" created by design engineering, operations departments, vendors, companies, and the need to accommodate wide variations in time and data availability in decision making. Long-held watchwords in manufacturing like "steady state" and "sustained production" are giving way to descriptors like "dynamic," "demand response," and "velocity," and a refocused interest in "agile." Vendor products in the Manufacturing Execution System space are giving way to more integrated operating tools under the concepts of Collaborative Production Management (CPM) and Manufacturing Operations Management (MOM). The ability to seek opportunity with *changing* products, markets, and manufacturing depends on innovation with products and materials and manufacturing operations and technologies. Smart Manufacturing recognizes that innovation, design, engineering, and manufacturing go together and feed off of each other. It broadens the base of innovation and innovators and has the wherewithal to be agile to the advantages of dynamics.

To be sure, Smart Manufacturing is intrinsically tied to Information Technology (IT)—certainly the IT of the past but critically going forward with the new wave of IT capability. This past decade alone has witnessed the rapid rise of data intensive informatics of unprecedented capability for vastly better understandings, and unimaginable new insights and new solutions at previously unthinkable scales. Past and present data from images, observations, and sensors, can be expressed digitally together to form databases of unprecedented scale, scope, and substance. Computational power and storage have reached "atomic" levels of fidelity in describing the fabric of physical, biological, behavioral, social, and humanistic expression. Mobile IT that is in the hands of much of the world's population has created yet-to-be realized data innovation and impact. Remarkably, these IT capabilities have come into being at a time when the problems of enterprise manufacturing interests and their solutions have become more complex, more inter- and transdisciplinary and with decidedly more integrative and involved partnerships (e.g., science and technology, operations and IT, workforce skills and jobs, business and operations, government and business).

These expansive new capabilities of networked-based IT can be argued both as a force to be dealt with as well as a transformative technology that offers new opportunities and approaches to deal with increasing complexity. Smart Manufacturing is the optimistic perspective that embraces IT as the marriage between physical and cyber technologies and workforce systems. Expansive application of enterprise modeling, simulation, and data analytics together with the existing installed base of serviceable control, automation, and business systems make the real-time actionable use of manufacturing data and knowledge possible. Enterprise management can move from transaction to performance management while control and automation can be extended and/or can be interfaced with complementary technologies to manage real time across many time scales. Data, algorithms, models, and simulations are rapidly becoming core innovation, planning, operating, and knowledge assets that are as valuable as the physical technologies. Smart Manufacturing encompasses physical, cyber, and workforce systems with an IT infrastructure that provides unprecedented access and flexibility.

As we proceed, we continue to build on the question that formed the basis of the SMLC back in 2007, "If Smart Manufacturing is so Smart, why isn't industry adopting it?" The consensus answer is multifaceted. Return on investment (ROI) is often a partnership with upstream or downstream departments, operations, or companies, and involves uncertainty with new business and operational models. Entry into Smart Manufacturing needs to reflect a wide diversity of technical, workforce, and investment readiness. Deployment of smart systems needs to overcome nonvalue infrastructure costs that can amount to over two-thirds of the total cost. Enterprise management needs to move from transaction management and companies need ways to test, evaluate, and drive down risk and build comfort with new technologies and business models. New smart technologies need to be retrofits of an installed base of serviceable control and automation systems and business systems. We are pointed to *infrastructure* that can support the alignment of people, product, and facilities operating in a next-generation level of dynamic and collaborative coherence with real-time, networked manufacturing intelligence applied digitally throughout the enterprise.

This book provides the underpinnings for understanding why Smart Manufacturing is so important and timely with respect to today's accelerating dynamic and enterprise sustainability, performance, and economic demands. Manufacturing and the application of manufacturing intelligence via IT are where they are today as a result of "first-generation" IT and the way business organized itself as a result of forces at that time. This book explains the foundational role of *time* historically and *why* control, automation optimization, and business systems evolved as they did. This perspective explains why individual manufacturing operations have been so extensively optimized into vertical "stovepipe" operations and why transaction-based management between operations has been adequate for enterprise management. The key insights from this book draw from an understanding of what has changed, why real-time enterprise (horizontal) management has become so important, and why it is so difficult. Most importantly, by taking this historical perspective, this book leads the reader step by step to an understanding of what is really meant by Smart Manufacturing and why in its expanded definition, Smart Manufacturing describes integrated capability and infrastructure that is essential to the evolution of twenty-first-century manufacturing.

> Quoting the Business Roundtable 2013 report, entitled Catalyst for Growth: America's Hybrid Infrastructure, "Infrastructure has always been important to our nation's economic growth and success, but the infrastructure needed for today's economy is rapidly changing with advances in IT. This new infrastructure—hybrid

> infrastructure—integrates both physical and digital qualities and is critical to delivering the next wave of innovation and economic growth in the United States. Successfully building out hybrid infrastructure will unlock new economic opportunities, create jobs and improve quality of life."

Twenty-first-century Smart Manufacturing is about the manufacturing sector's "hybrid infrastructure." Smart Manufacturing encompasses the public and private partnerships to build infrastructure that no one company can build and still support public- and private-sector interests. Smart Manufacturing hybrid infrastructure is comprehensive, attends to new workforce skills, reflects supportive access policies with attention to privacy and security, and addresses the value of shared data to industry and proprietary data to individual companies. It is itself both agile and flexible with access to commercial and public resources, assimilates new technologies in a market-driven fashion, and makes it possible to orchestrate actionable information among competing commercial interests. Smart Manufacturing is about ensuring that manufacturing can meet next-generation demands for environmental sustainability, safety, product value, productivity, and economic competitiveness.

Dr. James F. Davis is the Associate Vice Chancellor—Information Technology and a Professor in the Department of Chemical Engineering at UCLA. He is the University's CIO and is cofounder of the Smart Manufacturing Leadership Coalition.

Preface

Here is something you may not know. Almost all the productivity gains of the last 50 years have not come from better chemistry or better design. Yes, there have been advances in chemistry, in materials, even in biology. But these advances have not by themselves increased general productivity, nor have even better management and financial controls, nor the ubiquitous MBA graduates and sophisticated financing instruments. Not even entrepreneurship in the heyday of Silicon Valley created huge productivity advances. And certainly, the relocation of many business units offshore produced only a limited, and possibly wrongheaded, accounting improvement in productivity, which has declined as wages and working conditions in Asia and Latin America have improved.

These productivity gains have come from better automation and control of the processes, whether they are continuous, batch, hybrid, or discrete, manufacturing. The same advances in integrated circuits and electronics that brought us the PC and the smartphone also brought us intelligent sensors, networked devices, and database-driven control systems.

Much has been made of the push to make manufacturing sustainable or green and of the present trend toward moving manufacturing back to the developed countries of North America and Western Europe.

The secret to making manufacturing sustainable is also better control. It is better transparency and data continuity between the sensor and the enterprise. So why aren't the theories that have led to this amazing advance in productivity being used throughout the enterprise? Why aren't they being used above the plant level?

In this book, we will attempt to explain why this is so, and what we can do to gain an entirely new and higher level of productivity by extending these techniques horizontally throughout the supply chain and vertically throughout the enterprise.

The big problem is that there are multiple time bases used in manufacturing. Manufacturing process control is real time. Enterprise control is transactional and not real time. Just the data conversion systems required for the enterprise to have access to real-time data are daunting.

In this book, we are going to show you how enterprise control can be made real time. This will make your enterprise agile, responsive, and maximize financial gain—just the things we all need to survive in manufacturing in the twenty-first century.

If you work in automation or in plant management, this book will give you tools and ammunition to help you win support from your chief executive officer (CEO) to improve your profits. Surprisingly, this is harder than it sounds. We will show you a much easier path.

If you are an enterprise manager, this book will help you understand what the future of enterprises like yours is likely to be, and what you can do to help bring that future. We will show you what you have to do to make sure that your enterprise survives and thrives in the twenty-first century.

If you are a C-level executive, this book will show you why most of your profitability initiatives do not work, and how to use the principles of process control to manage and measure your business more effectively—in real time.

Business is out of control. We all know this, and we are all looking for a way to bring it back under control.

We are going to take you through a short history of why business is organized the way it is—from piecework and cost accounting to creating a transactional environment. And then we will discuss the driving forces that are pushing the enterprise to abandon that transactional environment and toward a real-time method for controlling business variables. We will discuss the implications of real-time business variables over a transactional environment, and we will show you how a process variable (like temperature) becomes a business variable. In order to do this, we are going to give you some control theory. We promise not to overdo it, but you have to see the link between control theory and the business of learning.

We are going to talk about how control theory is applied in the plant environment, and let you see how model predictive control bridges the gap between plant and enterprise control.

In order to understand what control theory has meant to enterprise productivity, we will take you on a tour of how control theory is used in the plant. We will talk about the actual discipline of control engineering and the development of real-time control. The first effort at real-time control was manual control, and as long as the processes were not overly complex, it was easily possible to operate a plant's functions manually. In the 1920s, Edgar Bristol Sr. pioneered the development of feedback loop controllers at the Foxboro Company. As these controllers achieved widespread use, feedback controllers became automated, additional and subsidiary loops were added, and control became complex indeed. Then model predictive control broke the boundaries of control and could be used in both the plant and the enterprise.

We will talk a lot about the theory of optimization and its applications both in the plant and the enterprise.

You also have to understand why the accounting systems that connect the work output of the plant to the enterprise are broken. Accounting systems now in common use were invented in the early 1800s to deal with piecework from skilled craftsmen. Then the Industrial Revolution turned unskilled and uneducated workers into cogs in the great machine of manufacturing.

There was a definite schism between white collar and blue collar, which is still visible in the system's divide between the plant (blue collar) and the enterprise (white collar). The successful drive in the twentieth century for better working conditions also made the "cogs" more expensive, and those costs were reflected in "productivity per employee" calculations. In order to make productivity gains, cogs needed to be replaced by automatic control.

Automatic control also required specialization. No longer could a "mechanic" work on every part of the plant. The plant systems fragmented into operations, maintenance, engineering, and management.

As training costs rose, and as cheaper labor in Asia and Latin America became available, downsizing came to the fore, and fewer control engineers did more with less. Automation engineering rather than instrumentation and control engineering became the discipline as the required domain expertise widened. Business leaders believed that they could do without high-paid engineering staff because the plant was built and running and stable.

Why did they think they could get away with that? They were interested in manufacturing commodities with stable business variables, and as long as things changed slowly enough, they could use the monthly financial roll ups (which ranged from 45 to 90 days after the fact) to operate the business. Unfortunately, with the rise of new competition in China, India, and elsewhere, there was no such thing as a stable enterprise. It became increasingly apparent that the problem was time. When you get the right information from your control systems, you also have to have it fast enough to matter. You can make a lot of the wrong product in 45–90 days, and trying to get information faster often leads to wrong information, and that leads to bad decisions and worse outcomes. Most business executives today get the wrong information at the wrong time because, while the control systems operate in real time, the enterprise business systems do not.

We will show you how business variables can be transformed with the inclusion of real-time process variables that are tied to financial metrics. When a system is stable, just like a process run by manual control, the enterprise can be run by manual control of business variables. Stable variables are slow to change and are easily controlled like a handwheel valve in the process. However, if you want your business to react in real time to changing economic and social conditions, you cannot do that without automatic control of the business variables. Control engineering is perfectly placed to solve this problem.

However, technology can be a barrier and a constraint. There are management barriers, silos, the cost of bridging disparate systems, and the need to fight the induced drag from unwilling labor and management who are allergic to changes.

So, what does an enterprise need to do to control itself? We have been looking, on the plant floor, at a tight integration between information technology (IT) and control systems. Many control systems use the same hardware and some of the same software as the enterprise does. The use of open standards and service-oriented architectures will enable us to inexpensively move information up and down the enterprise, vertically, and throughout the supply chain, horizontally. We can do this with highly integrated information and control architectures.

Efficiency at the plant level is not sufficient. Enterprise-level efficiency can result in too few plants, too few products, and no agility. It is actually possible to be efficient and not profitable. If we resort to just one control methodology, that of cascade control, we can trump efficiency alone and provide the right information at the right time to the right people for correct decisions. That way, decision support is used to support the making of decisions, and not as an "after action report" when the decision is seen to have been a failure.

In the future, we will see real-time profit control throughout the enterprise, with business-savvy control and automation professionals as the necessary talent base. We will do this with integrated systems that control the entire enterprise.

We would like to acknowledge and thank Carol Vallett for the hours she spent on reviewing and correcting the text and Melanie Russell for her technical reviews of the materials.

KEYWORDS

asset control, asset management, control theory, enterprise control, enterprise control systems, Industrial Revolution, industrial safety, industrialization, industrial business, industrial operations, operations empowerment, operations excellence, operations management, organizational silos, process control, profitability control, speed of business

1

A Retrospective

Industrial organizations have reached their current situation as the result of a long series of evolutionary steps. Today, there are remarkable similarities between the organizational structures of most major industrial businesses. People entering the industrial workforce may take existing organization structures and processes for granted without truly understanding how they evolved. When we merely accept things as they are without understanding the reasons and forces that shaped them, it can prevent us from really understanding how they may be changed to better meet the needs driving business today. For this reason, a concise discourse on the evolution of industrial organizations and operations may be useful.

Humans have been producing products throughout recorded history. The ability to manufacture things is one of the attributes that sets humans apart from the rest of the animal kingdom. Yet, we often begin discourses on manufacturing from the Industrial Revolution. But it is helpful to look at manufacturing prior to the Industrial Revolution since many of the characteristics of the Industrial Revolution were a direct result of manufacturing operations prior to industrialization. In a reasonably simplistic sense, prior to the advent of industrialization, most manufacturing was done by family-based, piecemeal manufacturing operations. Perhaps, the most visible example of this would be the manufacturing of textile goods. Textile manufacturing was also one of the first segments to be impacted by industrialization. Typically, a family that owned a farm producing cotton or sheep might realize some of the value of producing the cotton or raising the sheep by harvesting the cotton or wool and converting it into thread and yarn. They would then either sell the yarn or thread to a neighbor who would produce clothing or other goods, or they might produce those goods themselves. Once they satisfied their family's needs for clothing and other products, they might sell the remainder to the local general store, which would in turn sell the products to their clients.

Although this historical perspective may be obvious, there are certain characteristics of this manual production approach that are implicit in the process, but important to understand when evaluating the evolution of industrial organizations. First, the process was obviously extremely slow moving from today's perspective. There were subtle advantages to this slow pace. The first advantage was that accounting for the business value of the production was fairly simple and could be done on a piecemeal basis. (Rather than continually pointing out the differences between cotton production and wool production, the remainder of this discourse will focus on wool. A similar description of the cotton processes would reach the same basic conclusions.) The family that produced the wool could fairly easily determine if they could convert the wool to yarn and thread in a cost-effective manner, or whether it would be more profitable to sell the wool to a neighbor who might be more efficient, effective, and profitable at the conversion process. A similar analysis could be made with respect to converting the yarn and thread to final products. The production was so slow that good business decisions were readily apparent and fairly easily made. Accounting in this manual production environment was a very simple process.

The second advantage was that the tools required to execute the production process were fairly simple and inexpensive. Converting the wool into intermediate products might require a simple manual spinning wheel. Converting the intermediates (yarn and thread) into final products might require needles and other simple accessories. Acquiring and maintaining this equipment were inexpensive investments and certainly much simpler than investments necessary for today's manufacturing operations such as complex manufacturing equipment, maintenance systems, and operations management.

The third advantage of manual production of textiles was that the primary cost drivers were raw materials and labor. The raw material, in this example the wool, was produced on the farm and could be seen as a simple byproduct of the overall farming operation. The labor cost was the time of the family members. The energy required to operate the equipment was essentially human power. Therefore, the cost structure of these operations was much simpler than that of current manufacturing and production businesses.

The fourth advantage was that the market for the goods produced was typically very local.[1] The market dynamics were well understood and fairly stable over reasonably long periods of time. Therefore, the value of the cost of raw materials required for the manufacture of textiles and the value of both

[1] This is even so while evidence exists of wool from England being found in China, and silk from China found in large quantity in Europe. The producers sold the wool, wool intermediates, or clothes close to home and the merchants made the distribution chain as long as needed. The producers neither had nor wanted any control over distribution.

the intermediate products and end products were very well understood, and the resulting value of manufacture at each point in the value chain was well known.

The fifth and final advantage of manual production was that the children of the producers were brought up immersed in the manufacturing operation and often acquired the skill set required to support the production at an early age. In essence, this meant that there was a very high level of craftsmanship in these manual production operations with mostly everyone in the family understanding and knowing all aspects of the operation. This was a critical success factor for these family businesses. The more craftsmanship a family might develop, the larger the scope of the value chain associated with the production of textiles that might be executed within the family operation. The larger the scope of the value chain the family could keep in house, the more profitable they could be.

In retrospect, all of these characteristics are somewhat obvious and perhaps did not need stating at all. However, it is exactly these issues that presented daunting challenges to emerging industrial organizations as the Industrial Revolution took shape. Today's industrial companies are still coping with the early solutions to these challenges. Humans have a tendency to accept the approaches and solutions of the past as necessary for effective operation today. Therefore, we sometimes miss the fact that yesterday's solutions are not sufficient anymore. Industry is currently undergoing a huge transformation for which new ways of thinking and problem-solving are required. Unless we understand the impetus behind the historical solutions and approaches, it will be difficult to understand which solutions still apply and which need to change.

THE IMPACT OF INDUSTRIALIZATION

The reason the textile industry was selected as the basis for the discussion on industrial organization is that it was one of the first, if not the first, to be strongly impacted by industrialization. Many history books on the Industrial Revolution focus on the textile industry for this reason. The remnants of the impact of industrialization on textiles are still part of the culture through museums and old factories in older industrial sections of England and New England. The following description of industrialization in the textile industry is an oversimplification, but helps to set the stage for the purpose of understanding today's manufacturing operations and organizations.

Industrialization first impacted the textile industry as the result of some very clever inventions, such as power looms and automatic spinners, and latterly, the cotton gin, all designed to make the production of textile intermediates and finished products faster and more efficient. Clearly, families manufacturing products manually would have a difficult time competing with factories turning out products at much higher volumes and at lower costs per unit produced resulting from these new inventions. It was also very expensive to acquire the new machinery, which was prohibitive for most family-based producers. Typically, a single-family farm could not produce enough wool or cotton to keep the new machinery operating at the levels that would justify the purchase. Therefore, it was not unusual for a number of family-based producers to join forces to acquire a power loom and the associated equipment. If one of the families had a large enough building to support the equipment, that building might become the factory. If not, a new factory building might need to be built. The new factories were typically located as close to a major town or city as possible to reduce the time to market and tap into the local labor force. If the previous, farmer-based producers did not have the resources to build the factory, an outside industrialist might fill that role.

In any case, it became immediately apparent that the traditional approaches to organizing a family-based textile business would not suffice for the new industrial operations. For one thing, the speed of production increased markedly due to the new industrial machinery. Instead of producing a small quantity of intermediates or end products each day, these new manufacturing operations began producing vast quantities. Production moved too fast to be able to continue with traditional approaches, such as piecemeal accounting.

To realize the returns on the huge investments required to build, equip, and staff the new factories, there was great pressure to operate these factories for as high a production output as possible. New organizations tuned to continuous bulk production were required.

The complexity of these new operations increased significantly over the traditional manual production operations. While traditional production did not require any special knowledge to operate and maintain the equipment, the new machinery was extremely complex to operate and maintain. This forced factory management to develop specialty skill sets for the operation, maintenance, supervision, and management of the factories. No longer could these operations afford to have the same talent to both operate and maintain the equipment due to the bulk production and the specialty knowledge requirements. This led to silos of organization each doing its specific task. These silos had to be closely supervised to keep the operation running at some level of efficiency.

Material sourcing also became more variable than it had been. Instead of getting all materials from a single farm, these factories required vast quantities of raw stock from multiple farms. With multiple suppliers of the raw materials, the quality of the incoming materials was typically highly variable. Some of the farms may have been close to the factory and some may have been quite a distance away. Transportation costs for the raw materials had to be considered, and raw material sourcing issues increased both the complexity and cost variability of the operations. To deal with this complexity and simplify their accounting, many factories started paying material suppliers on a regular basis, perhaps monthly, for the material they supplied. Since the factories were the power base of the value chain they could essentially dictate the terms of how they did business. The suppliers basically had to comply to do business with the factories. Monthly billing and payment schemes evolved out of shear practicality. But one implication of going to monthly billing models was that accounting systems had to be adjusted to this time frame. Therefore, the practice of monthly accounting and reporting became a mainstay of these early industrial operations and is still the standard today.

The new industrial equipment required new sources of energy to operate. Early in the Industrial Revolution the new energy sources might be found by utilizing water power through water mills or other similar power sources, but as industrial demands increased electricity became a primary energy source for industrial factories. Once again, for practical reasons, the energy suppliers billed for the energy consumed on a monthly basis further supporting the concept of monthly accounting. The cost of energy became a major cost component of plants and factories that had not really existed prior to industrialization. Often the responsibility for electricity production was assumed by governments as a necessary infrastructure investment to encourage both industrialization and convenience for the general population. This meant that the cost per unit of energy consumed was initially highly regulated and was therefore fairly constant over long periods of time. The lack of cost variability was a bit of a blessing to industrial companies since they really did not have to worry about variability in the cost they paid for each unit of energy they consumed. If they could keep overall consumption of energy low, it would directly translate into a corresponding reduction in their electric bill. Nonetheless, the addition of energy cost to the overall cost model for industrial operations contributed to increased complexity of the cost accounting systems necessary to report results.

To exacerbate this difficult and complex situation, the market volatility of the newly produced goods tended to increase fairly significantly as more and more factories were developed. Over a very short period of time, new, lower-cost products started to become available. As new factories came into being, competition increased. Producers began to look for new markets, primarily by expanding the geographic distribution of the products produced. This not only created a new significant distribution cost in the value chain, which also increased product price variability, but it also increased the scope of the competitive market. Supply and demand issues increased in complexity. Product cycles also started decreasing with the combination of industrialization and resulting fashion trends. All in all, the complexity and new specialty skill sets and knowledge bases required to support the Industrial Revolution presented huge challenges in a variety of areas. One of the more challenging areas impacted was industrial organizations and labor management.

ORGANIZING FOR INDUSTRIALIZATION

The challenges associated with developing organizational approaches to deal with the new industrial organizations were every bit as daunting as any of the challenges associated with the Industrial Revolution. Industrial organizations needed to be developed to match the changing dynamics, skill sets, complexity, and difficulties associated with industrial production. The concept of the *manufacturing generalist* associated with manual production could not be sustained. Each aspect of industrial production required detailed specialty knowledge leading to a specialist-based organizational approach. Specialists were required to operate and maintain the new machinery, to design the manufacturing processes and continually improve production efficiency, to account for the business, to supervise across the operations to management supply and product chains, to plan and schedule production runs, and so many other functions required to effectively manage an industrial organization. Most of these functions did not previously exist, at least in the form required for industry, and had to be developed essentially from scratch in very short time frames. Some organizational approaches were adopted from military and business functions, but some of the organizational challenges associated with industrialization were unique and new and unprecedented approaches were required.

One area of organizational focus that actually went surprisingly well in the early stages of the Industrial Revolution was the basic and overall operation of the factory. The reason for this was that the initial workforce available to staff the operational aspects of the new factories typically came from the traditional family businesses and the talent base was comprised of craftsmen. That is, the initial workforce, at least in the textile industry, often grew up in family-run textile operations and essentially knew the trade inside and out. Although they may have had to learn how to operate the new machinery, most of the new machines were developed to automate functions that they already know how to do manually. As such, they truly understood what the machinery was supposed to accomplish and they tended to pick up the operations of these machines fairly quickly and easily. Visiting some of the old textile mills and historical sites it is striking to understand how quickly these craftsmen tended to adapt to the new industrial environments.

The biggest challenge in the operational area came when the number of factories started to grow to the point at which the craftsmanship-based workforce was stretched thin, and as some of the initial craftsmen started to retire and leave the workforce. At this point the available human resources to staff the labor force typically came from farming communities and were not craftsman-based as they had no longer been raised in the manufacturing environment. They were actually highly unskilled and uneducated. At this second phase of the Industrial Revolution there was a major management challenge involved with taking advantage of the available labor force in a reasonably efficient manner. The response to this challenge served to shape industry for decades. In many ways it still shapes attitudes and approaches to management of industrial labor forces today. For this reason this will be discussed in some detail a little further on.

Talent to maintain the machinery also presented a challenge to management. Although there had been blacksmiths and mechanics and other similar trades prior to industrialization, the new machinery was truly novel and considerable effort had to be invested in developing the skills necessary to keep the equipment operating. Maintenance teams were completely separate from operations teams out of pure necessity. Each skill required significant and unique capability. Further supporting the functional separation, the maintenance teams often had to perform their functions at the same time as the operations teams. The result was that, organizationally, operations and maintenance evolved to be completely distinct teams. To add to the challenge, operators of the equipment were typically charged with getting as much production out of the plant as possible, while maintenance was charged with making sure that the equipment was available (up and running) to operations. These two primary objectives tend to be at odds with each other. The harder the operators drive the equipment, the more likely it will break down. To increase the availability of the machinery, the maintenance teams needed access to the equipment and getting the necessary level of access typically required the machinery to be shutdown—which meant that the machinery was not available to operations. The result of this was that a natural distrust and even dislike developed between operations and maintenance departments, which often still persist today. This distrust has resulted in fairly non-cooperative organizations.

Financial accounting organizations formally developed during the early phases of the Industrial Revolution. Previously, the task of accounting for the value of the operation was fairly straightforward. Since the manual manufacturing operation proceeded at such a slow pace, the primary variables involved in determining the value of the production output were fairly simple and stable. Determining the value of the operation was equally simple and really did not require special skills or talent. It was merely part of the normal family-run business operations. With industrialization this changed drastically.

Production became so fast that accounting for each item as it was made, in a piecemeal fashion, became impractical. Many of the primary cost drivers of the operation, such as energy cost and feedstock cost, started to be billed on a monthly basis making accounting for the cost and profit of the operation essentially impossible within monthly boundaries. The result was a compromise. Monthly cost accounting and monthly closings of the company's books became standard practice. This compromise did not have much of a negative impact on the management of the operation primarily because most of the variables associated with cost and profit did not vary over the full monthly periods. For example, the price being paid for the electricity to power the manufacturing machines was typically government regulated and might not change at all for as long as a year. With this relative stability of business variables, the monthly accounting compromise worked reasonably well.

New organizations in the emerging industrial companies were developed to manage the volatility of the commercial markets into which the products were sold. Considerable competition evolved very quickly for the available market between industrial firms in relatively close geographic proximity. Merely producing products and providing them to the local general store was not a good enough approach for dealing with these competitive markets. New organizations focused on marketing, market analysis, and merchandizing and selling of products became essential to the success of industrial companies. Often the companies that most succeeded in the competitive marketplace were not those that manufactured the products best or even those that manufactured the best products, rather those companies that were most effective at sales and marketing. To a large extent the same is true today.

As industrialization started to mature it became clear that effectively designing manufacturing processes, and continually improving on those processes, was essential for both creating and maintaining a competitive edge and for operating the factories and plants in a safe and responsible manner. To meet this requirement new skills and talents needed to be developed specifically focused on these tasks. Engineering departments started to be created and evolved along these lines. Many colleges and universities developed mechanical, electrical, and chemical engineering programs to meet these critical needs. As this talent base became available in the industrial marketplace, the engineers were hired into industrial companies in unprecedented numbers.

Manufacturing processes became very complex very quickly. One of the important engineering disciplines required to effectively deal with this complexity was control engineering. Control engineers worked to apply control theory to bring complex industrial plants under tight control in order to efficiently produce products. The field of control engineering grew as industrialization grew. From the early 1900s through the 1980s the discipline of control engineering became a critically important discipline. However, as sophisticated control systems developed, control engineering declined to the point at which today it is almost a lost art. This is very unfortunate. As we will demonstrate, industry is currently undergoing one of the most significant transformations since the inception of the Industrial Revolution. This transformation is providing unprecedented challenges to industrial management. As it turns out, the discipline required to meet the challenges is based on control engineering. Many in industry believe that control engineering is in a decline that will essentially lead to its demise. We believe the heyday of the control engineer may be imminent.

Finally, to deal with all of these evolving organizational teams, overall plant and business management talent needed to be developed. In the early phases of industrialization this management talent was often provided by the entrepreneurs who invested in the plants and factories or, in the case of a consortium, by the lead investors, often who had substantial knowledge of the manufacturing processes in which they were investing. As the industrial organization evolved with more and more complexity, the need for a more sophisticated approach to industrial management became apparent. Colleges and universities began training potential managers with the cross-organizational understanding and business skills necessary to fill this demand. In areas in which formal training programs might not be available, industrial companies often put promising management talent through a rotational program through the different departments in their plants to provide them with the broad exposure required to understand the business. These up-and-coming managers might also be placed in an apprentice program under a senior and experienced manager to round out their preparation.

All in all, the emerging industrial businesses presented novel and complex organizational challenges that required completely new thought processes and approaches. Almost overnight multiple new functional organizations had to be developed in each plant and factory almost simultaneously. This was a tremendous challenge that the industrial leaders of the day met with great efficiency and effectiveness in general. Many of the organizational decisions and approaches made in these early days still impact industrial organizations today.

As new industrial drivers emerge industrial companies should continually question whether the organizational approaches developed over a century ago are still the most appropriate way to operate industrial businesses. Unfortunately, humans tend to be creatures of habit. Once an approach becomes accepted and has been working for a period of time, we tend to get very comfortable with it and not question its validity. Major new market drivers are starting to impact industrial businesses today that will require a reanalysis of decisions of the past. Industrial leaders are going to need to break out of their comfort zones to make the changes that most effectively manage these new drivers. Understanding the reasons that today's organizational approaches were developed should help leaders to understand how they might need to be adjusted.

THE SEPARATION OF OPERATIONAL MANAGEMENT FROM BUSINESS MANAGEMENT

One of the early and lasting impacts of industrialization was the development of two basic levels of management in industrial organizations—operational management and business management. The operational management team was tightly aligned with the engineering talent and responsible for the efficient operation of the manufacturing plants. The business management function was charged with managing the basic business functions of supply management, marketing, merchandizing, accounting, human resource management, and investor management of the overall industrial business. Like so many other aspects of organizational design in industrial companies, this two-tier management approach was highly practical with respect to the manufacturing environment. Industrial businesses, out of necessity, operated on two, very different time frames. The manufacturing processes managed by operations management teams were high speed, continually producing products, and had immediacy associated with their operation. They had to be effectively managed as the product was being produced. The business processes, on the other hand, managed by business management teams were much less immediate and essentially operated on a monthly time frame due to the monthly nature of the billing systems and accounting systems. Industrial organizations were designed around two time clocks—a real-time clock for operations and a monthly time clock for business.

The primary objective of business management was to manage and report on the profitability of the operation. Since many of the critical variables necessary to calculate profitability—such as energy cost, material cost, and production value—were billed on a monthly basis, it was not possible for business management to resolve the profitability of the operation any more frequently. Also, most of the external variables driving profitability, such as labor cost, raw material cost, energy cost, and product sales value, were highly stable for more than a month at a time. For these reasons, the primary nonoperational decisions required to manage the profitability of most industrial companies could be effectively made on a monthly or longer basis without negatively impacting the company's profitability.

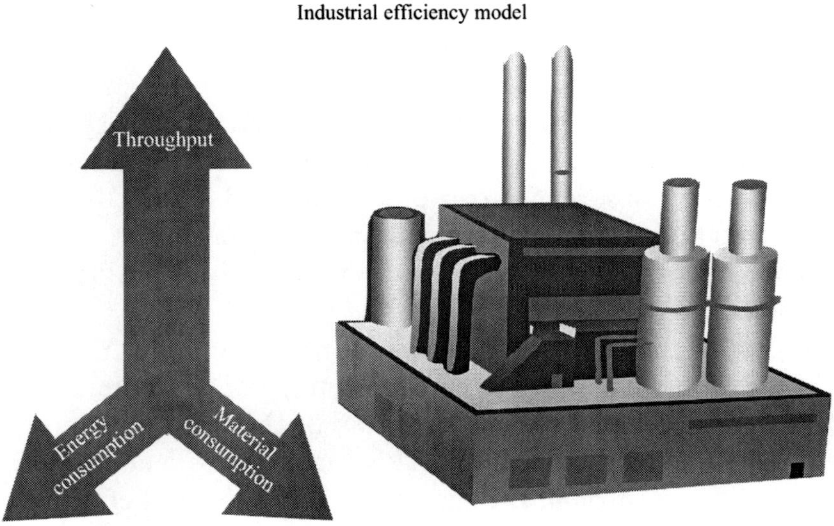

Figure 1.1. The traditional components of a process efficiency improvement strategy involve increasing the throughput of the process while minimizing energy and material consumption.

The primary objective of operations management was to manage the efficiency of the operation. The efficiency could be defined as producing the desired amount of product (throughput) at the lowest operational cost (Figure 1.1). Since the cost of energy, labor, and materials as well as the value of the products produced were all essentially fixed over full monthly periods of time, the objective of operations management could be simplified for meeting production volume requirements at minimal energy, material, and labor costs. If the industrial plants were efficiently operated by this definition then the profit from operations would be maximized. In other words, operations management did not need to focus on the profitability of the operation because maximizing efficiency would directly maximize profitability. Therefore, the entire perspective at the operations management level focused on maximizing the efficiency of the operation and the desired profitability was simply assumed to follow.

The efficient management of industrial operations presented daunting challenges in its own right. As industrialization continued to expand, manufacturing operations became more and more complex. In oil refineries, for example, producing the refined end products efficiently may require setting and maintaining the values of thousands of interrelated flows, levels, and temperatures to specific values based on the characteristics and attributes of the crude oil (feedstock) being processed. To early industrialists the concept of keeping all of these variables at the precise values necessary to produce the products efficiently must have appeared to be almost chaotic. However, problems, such as these, which may initially appear chaotic can be effectively solved by partitioning the overall problem into small components, solving each component and then coordinating across the components. This was initially accomplished using the labor force and the Scientific Management approach. Each operator would learn to control the variables in his or her assigned domain. If all of the operators across the plant were effectively controlling each of their assigned variables as specified, and if the engineers developed the correct specifications, refined end-products would result. The solution to this complicated operational management problem was the effective application of control theory. The problem itself was not a chaos problem—it was a control problem.

As the cost of the labor force increased over time, industrial companies found that they could not afford to have as many frontline laborers. Automatic controllers were developed to take on many of the basic functions of control previously done by the operators. This enabled experienced operators to supervise the process through the automatic controllers and effectively cover a much larger domain of operation. As this scope of operation increased, the experiential learning of the operators also increased. They began to view the processes in the plant from a real-time perspective that had not been previously possible. And they developed more and more experiential understanding of the operations in their domains.

Whether the actual mechanism for the control of the critical operational variables of the plant was human operators or automatic controllers, the separation of operational management from business management worked quite well as long as the business variables associated with the profitability of the operation were fairly stable for long periods of time. Considerable effort was expended through the financial accounting organizations to produce the information required to report the results of the business and to provide essential management decision support. The information required to effectively make the key business decisions to manage the profitability of the business was typically available to business managers in an appropriate time frame (monthly) to support business decisions. During the early phases of the Industrial Revolution, the information was produced manually by a number of cost accountants and bookkeepers. Over time, more and more sophisticated tools were introduced to automate much of the manual information management systems. Since the 1950s, digital computers were effectively deployed for this purpose. However, even with high-speed computer technology the standard for business information management was based on monthly reporting requirements.

It is interesting to note that the word "management" is most commonly used to describe the processes utilized at the business layer of industrial companies, while the word "control" is most common at the plant and factory levels. There is more meaning to this word usage than may initially be understood. We discussed this distinction with an executive of the world's leading

Enterprise Resource Planning software company, Sudipta Bhattacharya,[2] when he was with SAP, a number of years ago. Bhattacharya asked if I knew what the difference between management and control really means. He went on to say, "Management is when you cannot control something." The implication of his insight was that at the plant floor either humans or automation technology are implemented to specifically control each variable to its desired value, while at the business level, the executives cannot specifically control profitability so they do their best to try to manage it. Even early on during the Industrial Revolution when the business variables were very stable over extended time periods, maximizing profitability was not an exact science in the same vein as maximizing efficiency through process control in manufacturing operation. Although business managers would have loved to control their profitability, the means to accomplish this were just not available to them.

As management science increased in sophistication there was a drive to develop clear performance measures for all components of industrial organizations. The performance measures at the business management levels of the organizations were developed to reflect the profitability of the operation. Many of the business *performance* measures were standardized for consistent external reporting purposes so that investors could truly understand how profitable the company really was. It made good sense to align the business executives' performance to the externally reported metrics. But the measures of the profitability of the business typically did not extend down to the operational levels of the organization. Because of this, operational management often developed a separate set of operational measures intended to reveal the efficiency of each function in the operation. These operational measures are typically referred to as the key performance indicators (KPI) of the operation. The basis of this approach to operational performance measures was that if efficiency increased, profitability would directly increase as well. As long as this direct relationship between efficiency and profitability held true, traditional KPIs provided a reasonable set of measures for the performance of the operation that aligned well with the business performance measures.

THE IMPORTANCE OF ELECTRICITY

During the first Industrial Revolution, factories needed to be located close to the source of power. Mill towns, such as Patterson, New Jersey, were located near water power, where millraces could be harnessed to turn shafts, while take-offs from those shafts were used to power individual machines. With the

[2] Bhattacharya later went on to become President of Invensys Operations Management and then Chief Executive Officer at Adani Group.

invention of practical steam engines, factories needed to be located close to both water and coal.

The important thing to note is how factories were organized. In order to use either water or steam power, all of the machines had to be connected to drive shafts that used take-off belts to power the machine. Each machine needed to be aligned with other machines in a long line so that the drive shafts could maintain their efficiency. It was difficult to reorganize machines, or to change manufacturing processes, because in order to do that, the entire plant, or at least that specific production line had to be shut down. This had the end result of lengthening product life cycles, and reducing productivity growth.

Professor Paul A. David, of Stanford University, discussed in papers in 1989 and again in 1990[3] his argument that the dynamo was a disruptive invention in manufacturing, but that it required a long time and several preconditions before its ultimate force was felt.

We will talk quite a bit about electricity in this book, because it is one of the most important cost areas for a manufacturer, and instability in electricity costs (brought on by deregulation) has made it extremely difficult to maximize profitability because the cost of electricity varies so rapidly.

Professor David sees the dynamo (electric motor and drive) as the dividing line between the first and second Industrial Revolutions, and he said in 1990 that the computer would be the dividing line between the second and third Industrial Revolutions.

> In 1899 in the United States, electric lighting was being used in a mere 3 percent of all residences and in only 8 percent of urban dwelling units); the horsepower capacity of all (primary and secondary) electric motors installed in manufacturing establishments in the country represented less than 5 percent of factory mechanical drive. It would take another two decades, roughly speaking, for these aggregate measures of the extent of electrification to attain the 50% diffusion level.

David went on to note that "the history of electrification after 1900 . . . lends considerable plausibility to the 'regime transition thesis' " of Freeman and Perez. "They suggest that productivity growth has been sluggish and very well might remain so because the emergence and elaboration of a new techno-economic regime based on computer and communications . . . will, more than likely, be a protracted and historically contingent affair."

And so it has been, from 1990, when David published, to 2014, when this book is published. In 1990, only 10 percent of businesses used computers. In

[3] David, P. A. "The Dynamo and the Computer: An Historical Perspective on the Modern Productivity Paradox." *The American Economic Review* 80, no. 2 (May 1990), pp. 355–361. Papers and Proceedings of the Hundred and Second Annual Meeting of the American Economic Association, published by American Economic Association.

2013, close to 100 percent were. In 1990, only 2 percent of the world's business information had been digitized, according to David—while in 2013, at least 85% of it was available in digital form.

David remarks that the transformation of manufacturing by electricity was a long and protracted process, and it was not until close to the end of World War I, "when regulated regional utility rates for electricity were lowered substantially in relationship to the general price level"—in other words, electricity became considerably cheaper, and it became price-regulated and manufacturers were able to view electric power as a stable variable commodity.

David goes on to point out that

> the proximate source of the delay in the exploitation of the productivity improvement potential incipient in the dynamo revolution was, in large part, the slow pace of factory electrification. The latter, in turn, was attributable to the unprofitability of replacing still serviceable manufacturing plants embodying production technologies adapted to the old regime of mechanical power derived from water and steam.

In the second decade of the 21st century, we find that the potential productivity gains from moving to advanced manufacturing technologies are again slowed by the belief that replacing control systems installed in the 1970s and 1980s is unprofitable because the plants they control are still performing profitably.

Even then, often electric motors simply replaced water power or steam and were installed to run the same decades-old manufacturing systems that had originally been designed for water or steam. This ***group drive*** methodology was only gradually replaced by the ***unit drive*** system in which a single motor ran each machine independently.

As soon as unit drive systems became the norm (which did not happen until the late 1920s), it became possible to move to more modern manufacturing plant designs, including the more modern cell system. A unit could be taken out of service, redesigned, and reintroduced into service without shutting down the rest of the plant.

Another significant innovation brought about by electricity was the ability to operate the factory in round-the-clock shifts. Lighting the plant by electricity made it possible to double or triple the output of a plant, without a major expenditure after electric lights had been installed.

But it is important to remember that this is all in hindsight. The evolution of manufacturing brought about by electricity took approximately 50 years, and it was only later in the period that critical mass was achieved. And it was not until critical mass was achieved that the unexpected and synergistic effects—such as the growth of shift work—became apparent.

Since Professor David wrote in 1990, much has happened in the world of computers. First- and second-generation microprocessors have been succeeded by embedded controllers. Desktop personal computers have been superseded by laptops and tablets. The Internet exploded into the world, and the availability of data has become, in the words of Chris Stakutis and John Webster, inescapable.[4]

We are now at the same stage as the "vice president of electricity" was in the early days of electrified manufacturing. While we know that there has been an explosion of data, we are still using both manufacturing control systems and economic control systems that are not designed to use that data to the best advantage. We are still using "group control" instead of "unit control" systems.

[4] Stakutis, C. and J. G. Webster. *Inescapable Data: Harnessing the Power of Convergence*. Indianapolis, IN: IBM Press, 2005.

2

Out of Control

THE BUSINESS OF INDUSTRY GETTING OUT OF CONTROL

The business of industry is currently going through a major transformation, which has already been mentioned multiple times. The forces driving industry today may be creating as daunting a challenge to the managers of industrial companies as those of the initial aspects of the original Industrial Revolution. However, unlike the Industrial Revolution, in which the primary forcing functions of the transformation were highly visible: new industrial machinery, the forces behind today's challenges are much less visible and in many respects much more subtle. A research project undertaken by one of the authors, Dr. Peter G. Martin, over the last 20 years,[1] involving approximately two thousand executives from a variety of industrial companies in all major regions of the world revealed a number of forces driving significant changes in industrial operations. These driving forces have tended to suppress plant profitability by causing high-degree fluctuations in critical business variables over very short time periods. The result is that while industrial managers believe that the efficiency of their production processes may be well controlled, they are having great difficulty managing the profitability of their business.

Although there are currently a large number of forces driving industrial operations, the five major industrial driving forces related directly to this industrial business challenge will be specifically presented. In interviews, focus groups, and other executive interactions, it has been surprising how consistent the executive perspectives have been as related to all five of these forces currently driving industrial businesses. It is also important to point out that although these five industrial driving forces will be presented separately, they

[1] This study is proprietary to Invensys PLC and has not been published.

highly interact with each other to further increase the level of dynamic complexity driving industrial companies.

Driving Force 1—Industrial Business Going to Real Time Variability

The first major driving force identified in the executive research might also be the one with the greatest overall impact on industry today, and that is the transition of key business variables of industry from high stability over prolonged time periods toward real-time variability. When business variables demonstrate high stability over prolonged time periods they are referred to as "transactional variables" because their variability is often tied to human business transactions. When the variables approach fluctuating on time frames consistent with the physical and chemical time constants in the manufacturing processes, they are referred to as "real-time variables." From this point forward the terms "real time" and "transactional" will be used as a bit of shorthand to convey this sense of variability.

Underpinning the basis for the separation of business management from operational management was the fact that the business variables associated with the profitability of the operation were highly stable for prolonged time periods. In the automobile industry, it started with the price of oil and Detroit's inability to speed up their command–control loop. The Japanese got inside the Detroit command–control loops and were able to be more agile and bring out cars that had lower fuel requirements much faster than Detroit did.

In the energy-intensive industries it started with the deregulation of the electric power grids; the stability of some of the key business variables directly associated with the profitability of the business began to break down. When electricity had been highly regulated by governments the price consumers paid for a unit of electricity did not change for months at a time. But as governments throughout the world began to deregulate the production of electricity, increased competition in the market started to drive price swings. Over time, these price fluctuations became more and more frequent. As these pricing changes became more frequent it became clear that the frequent price changes would be very difficult for consumers; so local governments took action to provide some level of stability through a degree of regulation. Today, in the United States, the price consumers pay for electricity may change as frequently as every 15 minutes. In the United Kingdom electrical pricing can change as frequently as every 20 minutes. The net result on industrial consumers was that at least one business variable directly associated with the profitability of their business, which had previously been highly stable for months at a time, started experiencing near-real-time variability. The impact of this shift was initially very subtle. Wherein traditionally a reduction in the consumption of electricity would directly translate into a corresponding reduction a plant paid for the consumed electricity, this may not be the case any longer. For example, if the energy consumed on a Tuesday was 5% less than that consumed on Monday, but the bulk of the energy consumed on Tuesday was during high-price periods, the energy bill for Tuesday might actually be higher. This was a dynamic that industrial companies did not have to deal with previously.

The pricing variability in the electricity market had a domino effect on other critical business values impacting industrial companies. Any other energy source, such as natural gas, or raw material source, such as metals, which required electricity in its production processes started to experience a higher degree of pricing fluctuation as well. Today, in many areas of the United States, for example, the price consumers pay for natural gas may fluctuate as frequently as electricity pricing. The prices of many raw materials might not experience quite as frequent shifts, but are starting to exhibit multiple fluctuations daily. Also, the production values of the products produced in industrial operations are experiencing similar frequent pricing changes. A review of a copper price ticker on any financial television program shows prices changing multiple times a minute!

The good news is that not all variables associated with the profitability of industrial plants are moving toward real-time variability. For example, the cost of human labor tends to be much more stable than production value, energy costs, and material costs. As a matter of fact, currently, it is primarily only these three profitability components that do experience submonthly variations. Since the primary objective of any profit-based industrial site is maximizing the production value of the operation and energy and material costs tend to be secondary objectives, it can be somewhat helpful to display the real-time components of profitability as a simple vector model (Figure 2.1). This model is referred to as the real-time profitability model for industry, because it shows the relationship between the real-time components of the profitability of industrial operations. It is important to recognize, though, that this model is not a complete industrial profitability model because it does not incorporate the components of profitability that are still highly stable. Nonetheless, this model will prove useful for discussions on how industrial companies can deal with the new real-time business dynamics. The point is that the business of industry has been trending faster and faster over the last decade and a number of critical business variables are actually approaching real-time variability. The executives who identified this trend expressed considerable concern over how they might be able to deal with it. A number of executives expressed their concerns concisely when they commented, "their plants are in control but their profitability is out of control." Basically, they were indicating that the variability in their profitability seemed to be unpredictable and all over the map.

Figure 2.1. Industrial real-time profitability model. The three components of industrial profitability that demonstrate real-time variability are production value, energy costs, and material costs. The primary business objective is maximizing the production value, but production value must be balanced with energy and material costs to maximize profitability.

Driving Force 2—Plant Floor Focus on Profits

The second driving force identified during the executive research process is very tightly aligned with the first. Executives commented that their operations management teams had been very effective at applying control techniques to improve the efficiency of the industrial plants and factories, but that the executives need them to shift their approach to also help control profitability. Clearly, this is a direct consequence of the belief that their plants were in control while their profitability was out of control, but this executive perspective is also at odds with the traditional separation of business management from operations management as previously discussed.

Fundamentally, this driving force on industrial companies is that maximizing the efficiency of industrial plants and factories no longer directly translates into maximizing profitability.

One interesting aspect of the conversations with executives on this driving force was that in the language they used to convey the issue, they correctly identified the type or category of problem this driving force was creating—a control problem. This is the first important step in determining the correct approach to solving a problem effectively. Although this may seem obvious, it really has not been. There have been articles written that describe the apparent chaotic behavior associated with profitability in industrial companies as a chaos problem, and they have attempted to apply chaos theory to its solution. Although there are certainly aspects of industrial businesses that truly are chaotic and do require chaos theory–based solutions, this is not one. This is a control problem that requires the effective application of control theory to be solved.

Driving Force 3—Operational Empowerment

The third industrial driving force identified through the executive research is that the downsizing that has taken place in operations, maintenance, and engineering departments over the past 15 to 20 years has been destructive to the performance of the plants.

As a result, empowering the employees who remain in the operations to perform their duties more effectively and efficiently must become an important focus.

Many executives expressed the belief that downsizing had been a simple solution during the economic sluggishness and downturns of the recent past, but that they may have downsized beyond the appropriate levels. Some also mentioned that once they present a downsizing plan to a Company's leadership team and board of directors, it is very difficult to back off. Interestingly, automation systems were often economically justified based on the headcount reductions they would enable.

The executives seem to now believe that these same automation and information technologies have to be repurposed, not to replace people, but to empower them to perform their duties more effectively and efficiently. Although the initial focus in many of these discussions were on plant-level personnel, the executives actually believe additional empowerment with the information needed to make better business decisions is required at all levels of industrial organizations from frontline personnel right up to the chief executive officer. The implication is that as the business of industry continues to move faster, every person in industrial organizations will need to respond differently in some manner.

Driving Force 4—Increased Production Agility

The fourth industrial driving force that executives discussed is that an increased level of agility is necessary because of the increasing speeds in the industry marketplace. The primary constraint on meeting this requirement is that many existing plants and factories were initially designed in an era in which the markets were stable and would support as much production as the operations could produce. Therefore, many plants and factories were actually designed from inception to operate at full production output for 24 hours a day, 7 days a week, and for years at a time. Agility was seldom a consideration. The net result is that there are many manufacturing operations that are not inherently agile in a market demanding agility. Not surprisingly, a good example of this is found in the electric power industry in which decades ago electric plants were designed for continuous operation at maximum output because the demand for electricity had far exceeded the supply. Changing the output of these plants is a difficult and slow process that tends to diminish the performance of the plant equipment. One current solution to this problem is that the power companies have supplemented the traditional plants with a number of smaller and more agile plants. The large plants are used to provide a base level of electricity to meet the basic requirements of consumers, while the smaller plants are brought on line to handle demand peaks. Notice that this solution does not really involve trying to make the traditional plants more agile, but utilizes a combination of traditional and new, more agile assets across the power company's enterprise in a clever manner. This approach preserves the traditional power production assets that actually tend to provide electricity at lower cost than the newer and smaller plants, while providing the degree of agility required, improving profitability while meeting the fluctuating demands of the marketplace.

Driving Force 5—Increased Organizational Collaboration

The final industrial driving force discussed by the executives involved in the research project was the need to overcome the silos of organization resulting from the Industrial Revolution, and develop a more holistic organizational approach. As previously discussed, when industrialization initially took hold, the level of complexity associated with emerging high-speed, high-production factories and plants increased significantly. Different organizational functions—including operations, maintenance, engineering, accounting, supervision, and management—were developed to deal with the complexity. Organizational departments developed associated with these specialty functions and, unfortunately, in many cases, very little interaction tended to occur between different departments. Over the past few decades, it has been common to blame computer-based technologies for the silo effect with phrases such as "islands of automation" or "islands of information." Although in many instances these technical islands did exist, they existed to match the organizational silos that evolved out of the Industrial Revolution. A more apt description of the current situation may actually be "islands of organization."

In industrial organizations today there are often a number of different departments and talent bases that work in relative isolation from each other and do not tend to work together to solve problems. Most executives interviewed seemed to have a gut feeling that this isolationist approach could not be good for their business. In fact, many executives had read works from the "Father of Management Science," Dr. Peter Drucker of Stanford University, who frequently encouraged collaboration across talent silos by contending that most constructive innovation happens when two unlike disciplines get together to solve problems. Although the executives expressed their desire to overcome the silos of organization by encouraging the development of a collaborative approach across their enterprises, they were not sure how to make such an approach a reality.

These five industrial driving forces are currently causing havoc to industrial businesses. Many methods and approaches traditionally used to drive maximum profitability do not appear to be working anymore. And frustrated executives are in search of new approaches that will help to put them and their businesses back into some level of control. The good news is that most of the talent and tools exist to meet the daunting challenges resulting from these driving forces. An understanding of the nature of the problems to be addressed and the skill sets and technologies required to meet them is absolutely essential to the ultimate success of these industrial businesses.

REQUIREMENTS FOR DEALING WITH REAL-TIME BUSINESS VARIABLES

The traditional separation between business management and operations management in industrial organizations was based on a simple concept; getting the right information to the right people at the right time so that they could make good decisions. As industry moves forward with real-time business variables, this concept still holds. In the day of transactional business variables getting the right business information to the right people at the right time involved providing information on the basic cost and profit aspects of the operation to business decision makers in the upper echelons of the organization on a monthly basis. Since the business of industry was stable for months or longer at a time, this monthly information reasonably underpinned good business decision making. Therefore, Management Information Systems (MIS) designed to produce monthly business reports for financial reporting to the outside business world were typically sufficient for business decision support.

Conversely, decision making *within* the operations was typically tied to the speed of the manufacturing process—real time. This is because many of the decisions made in the plant, such as changing flows, speeds, temperatures, and pressures were made to directly impact and control the performance and efficiency of the process. Automation systems designed to support effective decision making in the manufacturing operations were designed to provide the right information—process values, production demands, and so forth—to the right people—operations, maintenance, engineering, and so forth—at the right time—real time.

In the emerging real-time business world a combination of the approaches is required to meet the organization's decision-making needs. The right information is both the business and operational information to support decisions, to the right person—the person responsible for making those decisions across the organization, at the right time—real time. Clearly, the traditional MIS and automation systems need to converge to meet this need. The combined information and automation system infrastructure will need to support business with an operational decision support across entire plants and enterprises as well as across the companies' supply and distribution chains. No longer can these be effectively supported by separate systems because they are all critical aspects of controlling the overall profitability of the business. The characteristics of these emerging systems need to incorporate some of the characteristics of the traditional MIS and automation systems. But one aspect that must go from the plant floor through the entire business operations will be the need for some real-time information at all levels. In other words, the future combined enterprise business control systems clearly need to be real-time systems.

BRINGING INDUSTRIAL BUSINESS BACK INTO CONTROL

The idea that many of the business components associated with industrial profitability are starting to experience real-time variability presents a daunting challenge to industrial business leaders. Many executives are at a complete loss as to how to meet this challenge. It is interesting, and in some respects a relief, to realize that most of the skills and tools necessary to rise to this challenge already exist in most industrial companies, although they may not have been deployed in an appropriate manner to meet the current challenge.

The class of problem that the current industrial driving forces are imposing on industry is a real-time control problem. Most industrial companies have great expertise in solving real-time control problems in their engineering departments. The basic operations of industrial factories and plants are typically controlled today through a combination of process control, logic control, and batch control strategies implemented by engineers and operated by operations staff. Overall, plant control strategies are typically very complex, often involving the coordinated control of thousands of process variables at the same time. Although on initial inspection, effective plant operational control may appear to be too complex to solve, control engineers approach the effective solution over this complex domain by breaking the problem down to small components—perhaps the control of a single process variable or machine. They then solve each component somewhat independently and then recombine the individual solutions into an overall coordinated solution using more advanced control and optimization strategies. Done effectively, the result is an efficiently operating plant.

The key to meeting today's challenges is to start applying that process control knowledge and toolset to control the business of industry. A detailed discussion of how control theory has been applied in manufacturing plants will be provided further on, followed by a presentation on how this same approach can be applied to real-time business variables. A critical success factor in rising to the challenge in industry is the recognition of what talent base in industrial organizations can provide solutions that meet the challenges. Many have intimated that the day of the control engineer is behind us. In fact, the pioneering work done by the great control engineers of the past has set the stage for controlling the profitability of industrial businesses in the emerging real-time world. We believe that a resurgence of control engineering is required and that industry must refocus on this area of engineering, which is rapidly becoming a lost art.

It is critically important to note that the focusing of control engineering on the business variables of industrial plants in no way implies that more advanced levels of control strategies are not required in the basic operational control of the plants and factories. In fact, for the control of the business to be effective the plant-level operational control must be as good as possible. In plant control theory, a control strategy called "cascade control" was developed decades ago to provide tighter control of process loops with particularly unstable components. A feedback control loop—the secondary loop—for the unstable component is first deployed to increase its stability. Then a higher level control loop—the primary loop—is implemented for the actual process variable of interest with the output of the higher level controller going to the lower level controller, resulting in a more stable overall operation. This cascade control approach will be more completely described in the chapter on control further on in this book.

The point is that the basic drivers behind the development of cascade control strategies in process operations are very similar to those of business control over process control. Considering the business control strategies as the primary loop of concern to the industrial business and the operational process control strategies as the secondary loop provides very important analogy. The control engineering approaches for traditional process cascade control strategies can almost be directly translated to business control (Figure 2.2). These concepts

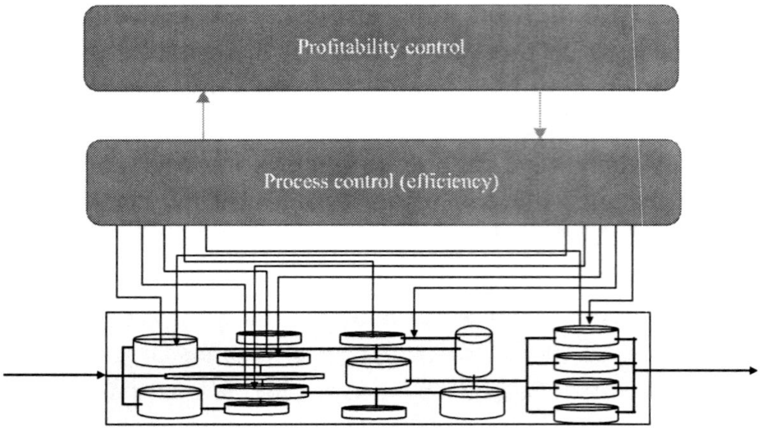

Figure 2.2. Controlling profitability in real time. The two fundamental control layers in industrial business involve controlling the efficiency of the operation through effective process control and controlling the profitability of the operation. In fact, the overall control strategy can be understood as a cascade control strategy in which profitability is cascaded to efficiency.

will be detailed further on, but understanding that the knowledge developed over decades of plant operations can be almost directly applied to meet today's challenges is a good starting point for effective solution approaches.

IT IS ALL ABOUT THE DATA!

In order to control the plant using business variables as well as operational variables, it is critical to have data. However, as Paul David explained about the dynamo, it is not enough to have data. It must be available to the enterprise, and the enterprise must have the tools to work with the data, in real time. David wrote that the computer has become ubiquitous, and so has data, as Stakutis and Webster revealed in their book *Inescapable Data: Harnessing the Power of Convergence.*

The problem is that the data are not getting from the plant floor to the business systems in real time. Manufacturing is about data, not about the machines. In fact, manufacturing theory and systems are technology independent. Companies have instituted Just In Time (JIT) procurement and inventory control systems entirely on paper using Kanban cards. From 1995 to 2000, one of the authors (Walt Boyes) worked for just such a company. Using such simple procedures, the company was able to reduce inventory by approximately 75%, speed up average time to manufacture by 50%, and improve profitability by

15%, entirely without a massive Enterprise Resource Planning (ERP) system, which we could not have afforded to install in any case.

In the 1990s, it was still not clear how information would be transmitted around the enterprise. Now we know. TCP/IP, Ethernet, and the World Wide Web have won over all other methods. Even the ubiquitous field device protocols like HART, Profibus, and Foundation fieldbus are run-through gateways into the world of Ethernet.

In 1990, individual PCs were used to control functions on the plant floor. Now a typical integrated operations center has perhaps three or four blade servers running as many as 20 virtual machines. Data historians and advanced control algorithm analysis are running in Cloud-based servers.

Not very many people know this, but there actually is an industry standard that governs the "duplex convenience receptacle"—the wall plug from which we get electricity. Electricity is so standardized and so available that no one thinks twice about compatibility. Stick your plug in and go. By 2014, data availability had reached the same level of standardization. OPC and OPC-UA were developed to be universal data converters in the same way that switching power supplies made electric plugs easy and standard.

It is easy to see that we have solved half of the data transmission problem. We can get data from the plant floor to the operations management layer in near real time. However, we still cannot get even operational changes back to the process control system in anything like real time. We absolutely cannot get real-time data into the transactionally based ERP system, let alone get the real-time control data back from the business system to the production system and thence to the control system. Manufacturing may be about data, but how do we get that data and turn it into information in a way that works?

Business decisions should be informed with both business and operational uncertainty and risks. Plant operations should be able to dial in along the spectrum of risk. Grade and product transitions should be optimized for situational economic, environmental, and energy impacts. Operational decisions should be informed by risk instability. Responses to abnormal situations should be informed by risk and uncertainty in operating limits and constraints.

Manufacturing ought to be safe. Working safely is more profitable than working unsafely and is much more economical.

Manufacturing ought to be smart, too. The data that are being continuously generated by smart machines and transmitters must be translated into actionable information.

Manufacturing ought to be responsible. We all live on the same planet. Manufacturing also ought to be sustainable. Energy and waste reduction savings go straight to the bottom line.

It has become clear, however, that smart manufacturing is not going to be created by automation and information technology (IT) vendors, or by end users in manufacturing acting alone. One of the reasons for the delay in adapting to smart manufacturing can be seen as analogous to the delay in the adoption of electrified manufacturing—the price of electricity had to come down, and similarly the price of delivering it, making it available, and using it in the plant. It may be that no single company, end user, or vendor will be able to bear the cost of developing the required systems and initiatives needed to give smart manufacturing the push over top dead center that will give it the momentum to transform business and manufacturing systems.

So just what makes manufacturing "smart"?

"We define smart manufacturing," says the Smart Manufacturing Leadership Coalition (SMCL),[2] "as an integrated, knowledge-rich enterprise in which all operating actions are determined and executed proactively applying the best possible information and a wide range of performance metrics."

SMLC[3] has created a list of 10 attributes of smart manufacturing:

1. Intelligent actions and responses
2. Operating assets are integrated and self-aware
3. Adapts to abnormal situation
4. Provides data when, where, and in the form needed
5. Proactive failure prevention
6. Rapid response for proactive control
7. Environmentally sustainable
8. People are knowledgeable, empowered, and trained
9. Smart manufacturing recognizes the limits of automation
10. Smart manufacturing drives strategic enterprise performance

As we have seen, in order to create such an enterprise-controlled manufacturing system, cyber infrastructure is key. Data interoperability must be maintained. The sensors that measure the processes must be networked. There must be a physics- and math-based understanding of material properties. Multiscale dynamic modeling and simulation must be core attributes of the system. Large-scale optimization must be applied, not just to the processes themselves, but to unit scale and plant-wide systems. In order to keep the

[2] The Smart Manufacturing Leadership Coalition can be found at www.smartmanufacturingcoalition.org

[3] In 2013, DOE EERE (Department of Energy Office of Energy Efficiency and Renewable Energy) announced that the SMLC won a 2013 Clean Energy Manufacturing contract to start developing the nation's first open smart manufacturing technology platform for collaborative, industrial, and networked information applications. The overall $10 million project will receive $7.8 million in funding from DOE EERE.

system working, all of this must be constructed on a clearly defined, scalable, requirement-based multilevel security system.

The path to twenty-first-century manufacturing, like twentieth-century manufacturing, leads through the development of new theory and practice. In the rest of this book, we will be producing some detailed, real-world examples of how this can be done, and how to put what we have learned into practice.

3

The Evolution of the Labor Force

Perhaps one of the greatest opportunities for improvement in manufacturing and production operations available is better utilization of the frontline plant and factory workers. To a great extent this unique talent base is highly underutilized. There is a very strong cultural bias against this talent base. Managers and engineers typically view the frontline workforce as a "necessary evil" within the organization. As the educational level and skill set of frontline workers has continually increased over the past decades, the prejudice that minimizes both their value and potential contribution continues to plague industry. Perhaps, it would be instructive to investigate the origins of this prejudice and the evolution of the workforce.

As has already been pointed out, during the period immediately prior to the onset of the Industrial Revolution, manufacture of goods was typically a family affair.[1] In textiles, for example, many of the goods were produced right on the farms that produced the raw materials—wool.[2] The farmers' wife and children might shear the wool, spin it into yarn, and even knit it into garments to be sold at the local grocery store. Since the children of these families essentially grew up in close proximity to the production process, often performing many of the production roles themselves, they developed a fairly high level of craftsmanship. Over time, they developed a strong, practical understanding of

[1] "Family" included the master–apprentice relationship, where the apprentice normally lived with and was a part of the master's family; as well as the remaining guild structures from the late Middle Ages and Renaissance. The meme of these relationships was imparting the knowledge of the family/clan/guild to a new potential member, not a hired person–employer relationship.

[2] To see an example of this persisting to the present day, visit Boyne Valley Wools, a mixed-use farm where Alice Fullam produces wool, spins thread, and produces garments for sale at the farm, which is located about 50 meters from the Bru na Boine visitor's center (the prehistoric Newgrange monument). http://www.boynevalleywools.com

the processes involved in producing the goods. Since this type of craftsmanship was developed in a large number of family operations during the initial phases of industrialization, there was a ready workforce available to the manufacturing and production operations with at least some level of the knowledge required to be effective operators.[3] As factories formed, industrial leaders found that in many cases this craftsmanship-based talent could build on their knowledge to learn how to operate the production machinery in factories fairly quickly and effectively. There were definitely challenges in learning the new machinery, but understanding the intent of these machines with respect to the process of producing products certainly contributed toward getting this talent base and the manufacturing operations up and running in fairly short order. It was not unusual for the managers of these plants to take extended time away from the plants since they knew that the workforce in the plants had the skills and knowledge to carry on without them for extended durations.

THE LABOR PROBLEM

The real labor problems started to arise when this initial talent base had been completely utilized and new factories and plants continued to come on line. The demand for people to work in these factories and plants rose exponentially. The available human resources (HR) to fill these emerging jobs did not have the same experience-based craftsmanship. In fact, much of the available human resource base had been brought up in agricultural environments—on farms—and with the manufacture of goods removed from the farms to local factories, they did not have the same opportunity to develop the manufacturing craftsmanship and knowledge found in the initial workforce. The majority of this available *labor force* was very poorly educated. Many could not even read, write, or do basic arithmetic.

Compounding this situation, the work structure on farms was very different from that required for manufacturing. Farms did not work off the type of structured timing system required in manufacturing operations. On a farm, the specific timing of the day's operations was not critical. The start of the day was more likely to be determined by sunrise rather than a clock. If a cow was not milked for 15 minutes or if the eggs were collected from the coops at 6:00 am rather than 5:50 am, it was not a major problem. Workers on farms developed a far different type of discipline from what was required in factories. They worked hard, but farm work is not nearly as regimented as factory work had to be. Work is done in spurts, not in a continuous manner. For example, harvest and planting are "all hands on deck" operations, but in Winter, there is much less to do.

[3] At Boyne Valley Wools, Fullam's two sons are very active in the business and have been since boyhood.

In a factory, if all the laborers did not show up for work at the same time, the factory may not be able to start up and run. Every minute required waiting for a late person or group of people was a lost opportunity to make money, which was unacceptable. On farms, workers could take breaks as and when they needed. In factories this was not the case. A machine operating without supervision for even 15 minutes might lead to a disaster. The traditional agricultural disciplines did not translate well into manufacturing operations. Essentially, the workforce available to fill the growing number of factory jobs was inexperienced, unskilled, uneducated, and even undisciplined—at least from a manufacturing discipline perspective.

On the other hand, there were huge numbers of people in this unskilled, uneducated workforce looking for work and they were available at remarkably low cost. As populations grew, farms could not financially afford to provide jobs for all of the available people. This meant that for industry there was a huge available workforce willing to fill positions in the factories, but the available workforce did not have the necessary skills, discipline, and education to perform the required work. Industrial management faced a very difficult problem.

The solution to the unskilled, uneducated labor force challenge was developed by an early and brilliant management scientist—Frederick Taylor. He termed his new approach *Scientific Management*. It was truly a great success. Without going into all of the details of Taylor's approach, he essentially suggested isolating a specific set of tasks that each person must complete for the manufacturing operation to work, train each worker to perform those specific tasks, and organize teams of workers in such a manner that if each worker did what he or she was supposed to do, the products would be successfully produced. In today's vernacular, Scientific Management essentially involved turning each laborer into a "robot"[4] by having him or her do a bounded piece of the overall work in a highly repetitive manner.

PROFITABILITY JUST HAPPENED

Over time, engineers in the industrial firms focused on how to improve the efficiency of each "work cell" and thereby slowly increasing the efficiency of the overall manufacturing process. It is important to note that even very early during the Industrial Revolution the focus of the frontline workforce within plants and factories was on operating efficiency—not profitability. The basic premise was that the more efficient a process was, the more profitable it would be. There was, therefore, no need to worry about profitability. At an early point this direct relationship between manufacturing and production efficiency to

[4] The word "robot" itself is from the Czech word for "worker."

profitability was certainly the case. The workers were trained to perform their duties quickly and correctly and in doing so, profitability just happened.

Implementing Scientific Management required huge numbers of laborers. Huge numbers were available and they were very inexpensive. Scientific Management was both a necessary and practical management approach for dealing with the labor force in these early industrial environments. Scientific Management was also helpful in addressing the challenges associated with the different personal discipline styles required for success in industrial operations as compared to that required in agriculture. Slowly the discipline required for the industrial operations became instilled into the laborers. This happened primarily by training laborers on very specific parts of the manufacturing processes, providing them with some understanding of how their job was essential to the success of the overall manufacturing plant, and with a huge amount of supervision and disciplinary oversight.

Early control panel in oil refining operation
courtesy of Foxborough historical commission

Industrial controllers were distributed throughout the operations, but control panels allowed multiple controllers to be mounted together enabling an operator to oversee the control of multiple control loops. Courtesy: Foxborough Historical Commission.

In short order, a huge unskilled, uneducated labor force began to assume the frontline operations and maintenance roles in industrial plants. The interesting thing about Scientific Management is that once the unskilled, uneducated labor force was trained and up to speed, they actually did a surprisingly effective job. In process manufacturing operations, such as oil refineries, chemical plants, and paper mills for example, each operator may have been responsible for controlling a small number of process variables (flows, levels, temperatures, and pressures). They would do this by viewing gauges that displayed the current values of the process variables and adjusting valves manually to adjust those variables to the prescribed values or set points. If operators could not read, the gauge might merely have green, yellow, and red sectors marked on the faceplate to indicate to the operator where the value of the variable should be maintained. Typically a green sector was provided for the desired area, yellow as a warning that the variable is moving to an undesired value and red to indicate danger. Operators would be taught to adjust the hand valves to keep all variables within their domains of control in the green sector. If all of the operators in the plants performed this action effectively, the products would be produced correctly. Getting the operators to control a small number of process variables was a simple enough task, but after operators got some experience, it was found that they might be able to expand their scope of control responsibility, which provided each operator with an expanded view of the production process. It also reduced the total number of operators required across the plant. With time and experience and the expanded view of the process, the operators began to learn about relationships and interactions within the process that may not have been previously observed, recognized, and understood. This growth of experiential knowledge started to increase the value the operators were able to provide. However, this increased value was seldom recognized by the engineers and managers in the plants who still perceived the frontline workforce negatively.

Operator view of early industrial plant
courtesy of Foxborough historical commission

Remotely located control panels increased the efficiency and effectiveness of operators. Not only could they oversee multiple process loops but they could visualize interactions between processes. Courtesy: Foxborough Historical Commission.

Historical records on the Industrial Revolution often present the perception that the early labor force did not operate the plants, factories, and mills very efficiently and effectively. Actually, there is considerable evidence that the opposite may have been the case. As operators gained experience operating the manufacturing processes, they often actually became very proficient at executing their responsibilities and also learned to effectively respond to unusual plant upsets and other unexpected operating conditions. In many cases, they learned how to respond to make the plant operate efficiently even when the engineers in the plant were not quite sure what or how the operators accomplished this. Engineers had developed considerable credentials academically in order to get their jobs. Academic credentials were very highly valued in the early industrial plants just as they are today. However, with the operators, a very different kind of education was taking place. The value of experience-based learning really began to have a huge positive impact. This was actually one of the earliest areas in which the huge impact and value of experiential educational approaches, which modern education theories are coming to accept and encourage, were realized. Many industrial organizations took advantage of this by developing "on-the-job" training programs to capitalize on the value of experiential training by developing the experiential training approaches for their labor force in a more systematic manner.

THE ORGANIZATIONAL SCHISM

A huge challenge associated with having to effectively utilize the unskilled, uneducated labor force using Scientific Management approaches was that an organizational schism developed between the labor force and the management (including professionals) in industrial organizations. Engineers were charged with the responsibility of designing, maintaining, and improving the efficiency of the manufacturing processes and operations. Much to the engineers chagrin, the labor force had to be inextricably incorporated into the manufacturing processes for them to work. Therefore, to a large degree, it became the primary responsibility of engineering to effectively deploy the Scientific Management approaches through the laborers in the plants while ensuring that the plants operated efficiently and safely. The general attitude of management and engineering toward the labor force was one of disdain and distrust. The lack of skills, education, discipline, and status that the laborers brought with them, coupled with the need for high-speed production and the inherent safety risks, resulted in engineers viewing their responsibility as actually "protecting" the plant from the irresponsible and untrustworthy laborers. Engineers limited the decision rights of the laborers, clamped the ranges that operators might operate the valves, and taught the laborers to limit their actions to small process adjustments and only when absolutely necessary.

Controllers at process vessels
courtesy of Foxborough historical commission

The controllers in this photograph are distributed across the process vessels. Operators could view both the controllers and get simultaneous sensory feedback from the process. Courtesy: Foxborough Historical Commission.

Engineers worked to minimize operators' involvement in the process. This reaction by engineers toward the operators led to the development of an operational philosophy and methodology commonly referred to as "operation by exception," which is still prevalent in production operations today. The operation by exception approach is actually designed into automation and control systems currently deployed in industrial plants. Operation by exception essentially implies that the operators are not to touch anything unless an alarm condition (exception) is identified. At that point they are to take care of the alarm, and once the alarm condition is resolved, go back to not touching anything. Many frontline operators in today's manufacturing plants have years of experience and truly understand the process from a perspective that no engineer or manager ever will, simply because they have watched the process operate for years. Education theory is starting to conclude that experience-based knowledge of the type found in experienced operators is typically more effective than academic knowledge learned in the classroom. Industry today has a huge experientially trained workforce that is still being treated like the unskilled uneducated workforce of the early Industrial Revolution. This is one of the major challenges and opportunities faced in industry. It is not a technology challenge—it is a cultural challenge that will be addressed more thoroughly further on.

Nonetheless, the schism that formed between management and labor had significant and enduring consequences. A pseudo-class system developed in industrial organizations with management on top and laborers on the bottom. Laborers worked long hard shifts at meager wages with very little organizational protection while management seemed to make more and more money on the efforts of the labor force. Just to make ends meet, often entire families worked at industrial labor positions with very poor working conditions and at extremely poor wages. Even young children entered the labor force often doing menial and often dangerous tasks. As the schism between management and the labor force widened as laborers saw little hope for advancement, the pressure in industrial organizations and in society as a whole grew. The people comprising the labor force grew in desperation over their seemingly hopeless situation.

Finally, over time, the labor force started to realize that they had power as a result of the vast numbers. There were many, many more laborers than managers. This power started to realize itself in very different ways in different geographies and across different political systems. Labor organized! In democratic countries trade unions started to form and the sheer number of laborers in these countries created an enormously powerful voting bloc. In other areas the same labor pressure led to socialistic and communistic movements. In either case, for the first time since the dawn of the Industrial Revolution, the laborers had power and a voice—at least with respect to governmental concerns. The result was major reform. New regulations were imposed on industrial companies with respect to age, working conditions, safety, and wages.

CLOSING THE SCHISM

The labor force and trade unions also made a huge push in support of better, state-supported public education from kindergarten right up through the college level. By the end of World War I it was not unusual for people entering the labor force to be able to read, write, and do basic arithmetic. By the end of World War II the typical person entering the labor force might have a high school diploma. And over the next 50 years the children of laborers often were afforded, and earned college-level educations.

The cry of the laborer was that, "I may have to suffer as a laborer, but my children will be better off than me." As late as the 1950s and 1960s this was still a very strong sentiment and attitude among union laborers. I (Peter) recall my father, who was a hard-working union laborer, emphatically driving into me and all of my brothers and sisters as we were growing up that we were going to go to college and that there would be no further discussion on the subject. He wanted to ensure that we would have an easier life than he did. Eventually, through hard work and study he actually ended up working his own way up to an engineering position. I recall sitting with him in our kitchen as a teenager, and him clearly pointing out that I was going to go to college. It was just an accepted fact. I also recall him telling me that he was not going to pay for it because he could not afford to. I fortunately never questioned his wisdom and it certainly served me very well. The point is that his perspective truly represented the general attitude of the labor force. This attitude helped to drive up the educational levels in most, if not all, industrialized geographies.

I (Walt) also remember my father similarly. He graduated from the eighth grade, and worked his way up to an Industrial Engineering position, and finally a series of senior management positions at companies like FMC. He became a manufacturer's representative in the instrumentation business later in life.

The consequence of the labor movement was that over this 50-year period the labor force went through a huge transformation, from an unskilled uneducated workforce to a skilled and relatively highly educated workforce. The problem was that management and engineering of industrial organizations seem to have missed the transformation. Even though wage levels and conditions improved, the general negative attitude toward the frontline industrial workforce changed very little. Even today, the frontline operators are still perceived by management and engineering, even if subliminally, as a resource that is untrustworthy and that must be severely constrained to protect the plant from their incompetent actions.[5]

A good example of this attitude can be seen in the industrial fad of the late 1970s and early 1980s referred to as *lights-out manufacturing*. The concept of lights-out manufacturing was sponsored primarily by engineers and based on the idea that computer-based technology had advanced to the level at which the operators in industrial plants could be replaced by technology. The idea was that if no operators were required to run the plants, all of the lights could be turned off because there would be no people in the plants, and therefore no need for them. Although the lights-out manufacturing movement gained significant momentum, it never became a reality. Although many manufacturing operations could be very highly automated, computer-based systems tend not to be very adaptable as they can only respond as they are programmed to respond. Engineers soon learned that even though they might be able to

[5] From my own (Walt) experience of working in the automotive industry, it was clear that the mechanics and line workers actually knew more about how to make cars and trucks than the design engineers and manufacturing engineers did, but they were seldom listened to. Eric Flint, the best-selling science fiction writer, tells of working as a machinist at Lockheed's "Skunk Works" and sitting watching a lathe not cut any metal for 8 hours one shift because he had been forbidden to go talk to the engineers, and the drawing was wrong.

computerize all to the normal and "expected" manufacturing operations, they could not possibly anticipate all of the unexpected events that might occur. They found that they could not program the computers to cover all potential actions and activities and, therefore lights-out became impractical. Humans, even frontline operators, tend to be able to respond to a wide range of unexpected events by learning and drawing conclusions from their experiences. Eliminating the operators and going lights-out did not and could not happen. The real point of reviewing the lights-out manufacturing fad is that it was a relatively recent movement providing a good example of the lingering negative perception of management and engineering toward frontline personnel.

Reactor control room
courtesy of Foxborough historical commission

Controllers and control panel in a refinery. Control panels could be located in a relatively safe location in the plant because the controller could be separated from the process via either pneumatic or electronic analog communications.

Automation technology continued to develop through the first half of the twentieth century. Simple mechanical control devices were replaced by pneumatic measurement and control devices. One nice feature of the pneumatic devices was that the air signals could be transmitted over a distance. This enabled multiple controllers to be located in a single station so that a single operator could view and manage a fairly large scope of the operation and observe interactions in the processes. With the expanded view came an expanded experiential knowledge of how the operation actually worked. Operators started developing unique understanding of the plant from the expanded view afforded them.

As the pneumatic technology continued to develop and as electronic analog technology began to replace the pneumatics, signals could be sent to much longer distances without significant signal degradation. This feature led to the further consolidation of control stations into either centralized or distributed control rooms. In either case, the operators in these control rooms had a much more expanded view of the operations of the plants than had previously been available. A broader experiential understanding of the industrial operations by the control room operators was a natural consequence.[6] This experiential knowledge that evolved in the operation staff should have been highly valuable to the industrial plants, but the traditional prejudice against the operators as a labor force served to limit their utilization and decision rights. Engineers continued to see it as their job to protect the plant from these low-end laborers.

Reactor control room
courtesy of Foxborough historical commission

As control systems evolved, the controllers could be located at great distances from the processes enabling the development of control rooms in which small teams of operators could oversee the operation of large plant areas.

Certainly, by the late 1960s and early 1970s, as the digital computer was just starting to be used for plant floor automation in industrial plants, the labor

[6] Indicators in control rooms were often installed along a horizontal axis on the wall, with the set point indicator in the center of the gauge. When the operator walked into the room, by a simple scan of the indicators, he could see that the process was "lined out" because all of the pointers were in the center of the gauges forming a line across the wall. If one was "out of line," the operator could see and respond to the problem at once.

force had become much more skilled and much more educated than their predecessors had been. Computer-based control systems started to replace the pneumatic and electronic analog systems and computer workstations started to replace the control panels in control rooms. Although there were a number of very important innovations made available with the introduction of computer-based control systems, perhaps no advancement had more impact on the development of the labor force than the computer workstations. With these workstations, operators had views of the manufacturing and production processes from multiple perspectives never previously available. They had multiple different perspectives of the operation showing both the current process variables and historical trends of the process data for entire plant areas, and sometimes for entire plants. And they learned. They actually developed a remarkable experiential understanding of the process not previously available. Operators became increasingly more valuable to the organization.

To put this into a little bit of perspective I will share a personal experience I (Peter) had during the early 1980s when I was an automation project manager. I recall kicking off projects with initial project meetings in site conference rooms with the entire project teams that would be involved, including both operators and engineers. To get the project off to a reasonable start we would typically review the piping and instrumentation diagrams (P&IDs) for the area of the plant into which the automation system was to be installed. I recall multiple occasions in which, while reviewing the P&IDs for the project a question would be asked along the lines of, "if this temperature changes will it impact this pressure?" At which point, almost simultaneously I would hear, "yes"–"no".

I knew one of them had to be right. I would then ask, "Who said no?" An engineer might raise a hand and say that he had gone through the P&ID's in great detail and found no relationship between that temperature and that pressure. This was certainly a reasonable response. I would then ask, "Who said yes?" Inevitably, an operator would raise a hand and say something like, "I have never looked at a P&ID and hopefully never will, but I have watched this plant run for years and can guarantee that every time that temperature changes that pressure changes. I don't know why and I don't really care why, but I know that it does." And every time this happened, the operator was found to be correct. The experiential knowledge they gained over years of operation was starting to become invaluable to industrial operations, but was still underutilized.

A NEW KIND OF CRAFTSMANSHIP

The experiential knowledge developed by operators through automation systems can actually be thought of as the basis for a new type of craftsmanship. Dr. Shoshana Zuboff, in her classic work, *In the Age of the Smart Machine* referred to the developing knowledge base in the operations staff as the "informating" of these people. Interestingly, humans tend to learn through feedback control mechanisms. The concept of "trial and error" often comes up in learning system theory. Trial and error is actually a feedback control process. The immediate feedback provided to operators through computer-based automation systems serves as a very effective learning mechanism. Over many years, operators evolved from unskilled, uneducated laborers to "control craftsmen," then to "automation craftsmen" and to the "information craftsmen" of today. Yet, they are still, in many respects, treated as laborers. Looking at a typical chemical plant facility, the boiler operator must be licensed, as must the water and wastewater treatment operator, and the environmental manager. However, the control system operators who actually run the plant have no license requirement, because of this belief that they are "just labor." What a waste of a valuable resource.

Figure 3.1 provides a graphical summary of the evolution of the industrial labor force as perceived by management and professionals. Coming out of the initial phases of the Industrial Revolution, and the trade craftsmanship-based labor force started to dry up, manufacturers developed their labor base from the available highly uneducated, unskilled workforce. With rapid industrialization this uneducated, unskilled labor force continued to expand. Dealing with this labor force caused huge management problems. Frederick Taylor

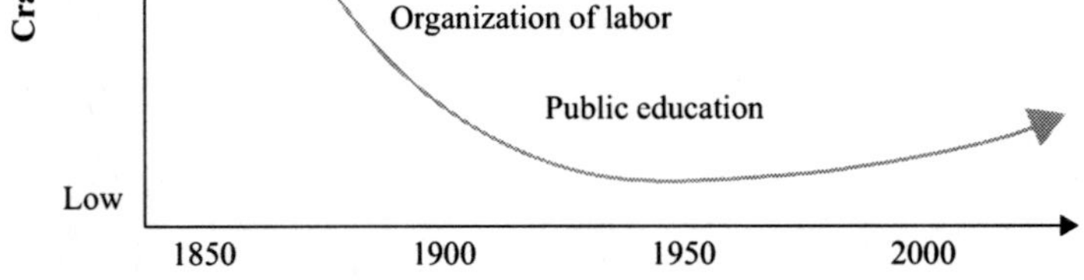

Figure 3.1. As industrialization took hold there was actually a high level of craftsmanship in the early labor force because craftsmen of small family businesses became operators in the factories. As the migration from farms to cities took hold, new factory workers were unskilled and uneducated and the level of craftsmanship declined. In democratic countries the laborers became a political force and pushed for programs like public education to improve the fate of their children.

developed a pragmatic management approach, Scientific Management, that proved successful in mobilizing the labor force, but also enforced a management perspective on industry essentially relegating the labor force to the status of necessary evil. As labor organized and became a significant political force, new governmental policies were developed, such as expanded public education, which fundamentally changed the nature of the available workforce. No longer were they uneducated as they once were, but in traditional industrialized geographies they are actually quite well educated. Yet, as the curve in graphic and current management practices shows, management and engineering still have a low view of the capability of the frontline workforce. It is this view that is a huge barrier to increased performance levels—not the workforce itself.

Figure 3.2 emphasizes this point. As new technologies were introduced into industrial operations to help control the manufacturing and production processes, there have been two major revolutions impacting the frontline workforce. The first revolution was triggered by the expanded view of the manufacturing processes provided by the expanded roles of the operators due to advancements in control and automation technologies. For this reason, this revolution is called the "Automation Revolution." As computer technologies were introduced to help control and automate manufacturing and production processes, they brought with them a true step change in visibility for the frontline workforce. The frontline operators went through an "Information Revolution," which enabled them to learn about the industrial operation from a perspective never before available. The combination of the

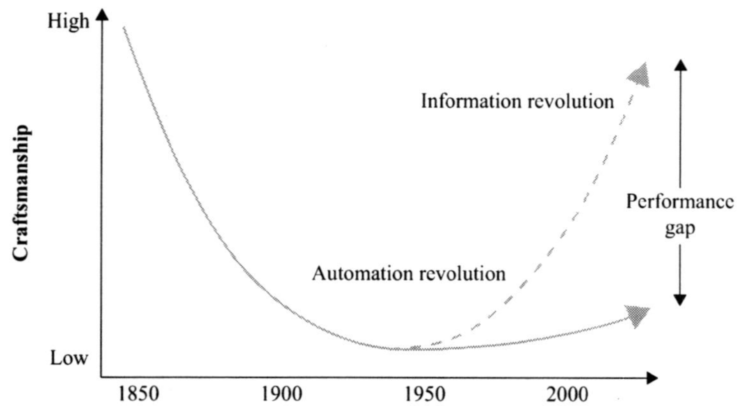

Figure 3.2. In parallel with the labor force becoming more educated, new technology advancement in the areas of automation and information technologies provided great opportunities for growth, but industrial organizations tended not to capitalize on it—creating a performance gap.

"We've found the key to productivity. It's Fred, down in the shop. He makes the stuff."

Figure 3.3. *Wall Street Journal* cartoon. This comic from the *Wall Street Journal* was intended to generate a laugh, perhaps because it is so true.

increased education levels of the incoming workforce and the increased visibility into the manufacturing processes, represented by the dashed curve, caused many of the frontline operators to become much more valuable to the operation. Unfortunately, management and engineering continued to perceive the frontline labor force in a traditional sense, represented by the solid curve. The difference between the two curves is the "performance gap" being suppressed in many industrial organizations.

Years ago, I cut this comic out of the *Wall Street Journal* because it represented the current management perspective of the industrial workforce so well (Figure 3.3). The comic is presumably a Board Meeting at which they are trying to evaluate how to improve the business. Almost in a surprised manner the chief executive officer is pointing to the frontline laborer—Fred—as the key. The reason this is so humorous is that it is so true! Industry has invested greatly in upper management and in professional talent, but has tended to ignore the frontline workforce, which is such a valuable resource and so ready to perform—if only allowed, enabled, and encouraged!

The entire mindset and vernacular around the labor force has to be replaced. Industry has to start thinking about the frontline personnel as managers of performance and treating them as such. The first step is for management and engineering to realize the value of this resource and shift their perspective from the workforce as being necessary evils to being valuable "performance managers." Once this barrier is overcome, the next step in transition is really

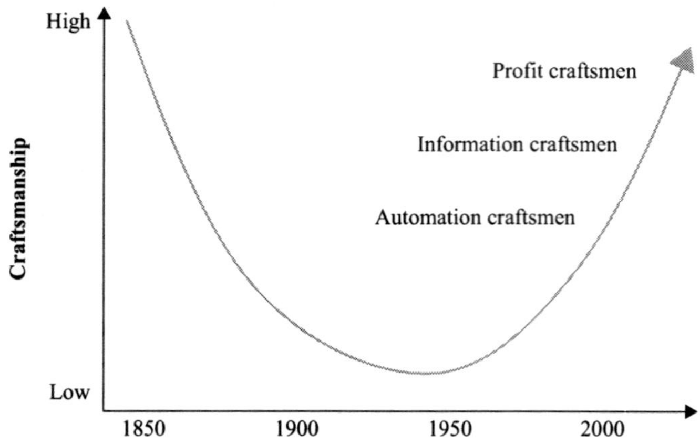

Figure 3.4. "The solution...laborer to profit craftsman." Transitioning the labor force to become profit managers is a natural next step in the evolution of the labor force if industrial cultures support it.

quite simple and straightforward, that is, to empower these performance managers with the information necessary to enable them to drive improved performance. The key to this is recognizing that the way manufacturing plant efficiency is maintained and improved is through the effective application of control theory. The way humans, including these performance managers, learn is through the effective application of control theory. The biggest performance problem in industry today, the inability for managers to effectively manage the profitability of their business due to the shift of industrial business variables toward real-time variability, requires the effective application of control theory. Therefore, providing the right information to these performance managers, in the correct context and in real time, will allow them to learn how to control and improve the profitability of the operation. As the graph in Figure 3.4 supports, developing and implementing an empowerment mindset and approach will help evolve the frontline workforce into automation craftsmen, then information craftsmen, and finally to the profit craftsmen that industrial companies need them to be. Simply stated, they will move from being laborers to performance managers.[7]

[7] It should be noted that human resources (HR) regulations in many companies draw a hard line between "managers" who are typically salaried and "labor" who are typically paid an hourly rate. In some companies, it is necessary for the "laborer" to quit and work elsewhere as a manager before the company will hire them back above the hard line. This is a hold-over enshrined from the time when there really was a class difference between managers (gentlemen) and laborers (commons). The hierarchical structure of many industrial facilities proceeds directly from this differentiation.

THE DOWNSIZING DEBACLE

The last 15 to 20 years have exacerbated the situation with the labor force as industrial economies and markets have gone through a number of recessions. When markets soften, the easiest way to respond is by cutting labor costs. Over the last 20 years, this downsizing or "rightsizing" of the workforce has impacted operations, maintenance, and engineering to a high degree. This has led to two critical challenges for industrial companies. First, many companies are operating at staffing levels that are below what is necessary to effectively drive performance improvements in their operations. Even if the remaining personnel have the right experience levels and skill sets to execute performance improving initiatives, it does not happen. As they are typically just too busy trying to complete their day-to-day tasks, there is no time to focus on performance improving initiatives. The result is that many industrial plants do not experience continuous improvement. In fact, they are not even operating as efficiently as they had been 10 years ago.

A number of industrial organizations tried to offset the downsizing by investing in technologies, such as automation systems, to supplement the personnel so that not as many would be required to run the plants. Certainly, there was some justification to this approach—at least until the downsizing impacted the engineering force, whose job had been to make sure that the automation technology was having the impact it should in improving performance. In many plants, the engineers just did not have the time to deal with expanding the capability of the installed technology. Automation systems were often installed plants to specifications that utilized only about 30%–40% of the available capability and capacity of the technology. The thought process was that the engineers would take advantage of the additional capability and capacity "later on," once the project was complete. Unfortunately, because the engineers did not have the necessary capacity, the expected later on never happened. The systems were often decommissioned 20 years after installation providing the exactly same level of functionality specified for the initial project.

The second critical challenge cause by the continual downsizing of the past 30 years has been the development of a huge age-based talent gap in many industrial organizations. As the downsizing of operations, maintenance and engineering progressed; very few new people were hired into industrial organizations. The result was that the average age of the workforce kept increasing. When I started in industrial automation in the 1970s, the average age of an engineer in industrial companies was my age, mid- to late-20s. Today, the average age is still my age—and I am obviously considerably older at this point. This is clearly shown in the demographic breakdown of International Society of Automation (ISA) members. The average age of ISA members is between 50 and 60 years, if retirees and life members are not counted. To make matters a bit more challenging, it takes a long time and considerable experience for a new operator, maintenance person, or engineer to come up to performance and to independently drive value. Today, there are many workers on the verge of retirement and a number of very young and inexperienced workers behind them—not quite ready to take over full responsibility. This has led to a serious experience gap that industry is trying to deal with.

Unfortunately, over the past 30 years, the attractiveness of industrial jobs in the minds of young people has declined significantly. Few people entering the workforce are inclined to take up industrial and manufacturing jobs. There has been an increase in business majors in colleges and universities and a decrease of interest in engineering. Industrial professions are not very attractive to young people. This further exacerbates the workforce challenges.

It is also quite interesting to consider generational differences between the retiring "Baby Boomers" and some of the younger generations. These generational differences provide both challenges and opportunities. Baby Boomers tended to be very detail oriented, wanting to understand how things worked right down to the lowest-level mechanisms, while younger generations tend to have a more abstract view of how things work but are very adept at dealing with newer technologies. Combining the thought process of the Baby Boomers with the technological comfort and agility of the younger generations can offer promise in bridging the gap. New approaches to education—both academic and experiential are required to help industrial companies meet this daunting challenge.

The most significant opportunity industrial companies have to drive improved performance and improved profitability is capitalizing on the HR in their organizations. A major part of the discussion in this section has been focused on the frontline operators and maintenance personnel, but engineering must also be included in any discussion of potential issues and solutions. The good news is that advances in automation and information technology, learning and educational theory, and management science have set the groundwork for a major leap forward. Many of the required pieces are in place. The remainder of this book will involve pulling them all together to create performance-based, real-time industrial enterprises.

4

Knocking Profit Out of Control

One important aspect of the early Industrial Revolution was that the piecemeal approach to the manufacture of handmade goods could not be sustained for factory- or plant-produced goods. With the advent of industrialization, goods started to be produced so rapidly that accounting for them a piece at a time was just not feasible. To exacerbate the problem of speed, some of the key consumables of production processes, such as energy and materials, could only practically be billed by their suppliers on a monthly basis. The net result of this situation was that monthly accounting became the fastest practical accounting approach industrial business could implement. The profitability of the plants or factories could be determined only as frequently as monthly as businesses "closed the books" on the month. Many of the most commonly accepted business practices of today, which are seldom even questioned as to their validity, were set into motion with the advent of the Industrial Revolution.

Closing the books takes time, too. Often, an enterprise does not issue its monthly accounting report for 10–20 days after the end of the month. This means that it takes longer than the month (30 days) to get the most current information.

It should be noted that the monthly cost and profit management introduced with industrialization was a compromise. Business managers would typically like to know how their business is doing on an ongoing basis—day-by-day or even hour-by-hour. The previous piecemeal accounting approach provided this type of visibility. Monthly accounting systems did not. They represent a compromise forced on industry by circumstances associated with industrialization.

One fascinating aspect of human behavior is that, outside of major forced transformations such as industrialization, humans tend to be extremely set in their ways and reluctant to even consider changing. This has certainly been the case with respect to monthly accounting systems. Over time, business schools started teaching the basics of monthly accounting systems and monthly business management, not as though they were compromises, but that they were the only way to operate businesses. When business professionals started earning college degrees with a focus on the workings of monthly business management, they tended to stop considering whether there may be a different and better approach.

Monthly business management cycles became the norm. In fact, monthly business management and accounting systems became so ingrained into the business culture and management thought processes that to even question the validity of this approach was almost considered taboo. This was supported by the reality that managing industrial businesses on a monthly basis actually worked quite well early in the industrial age and for most of the following decades. The reason was that the critical *business* variables associated with the manufacture and production of goods exhibited very little variation over very long time cycles—typically much longer than monthly. For example, the price industrial plants paid for their electricity typically was constant for a year at a time. Measuring the energy price on a monthly basis resulted in little or no loss of information. The same was true for other critical business variables, such as labor costs, raw material costs, and production value. With the variables of the business of industry changing so little and so slowly, measuring them monthly did not inhibit reasonable business decision making, and proved to be quite adequate.

Contrast this to the timing required to operate manufacturing and production processes. The key variables associated with the effective management and control of industrial processes—variables such as flow, level, temperature, pressure, and speed—change in time frames associated with the natural time constants dictated by physical and chemical laws—that is real time. Pressures, for example, can fluctuate within time frames of less than a single second. If industrial companies were constrained to the point at which they could only measure each of these variables once a month, the manufacturing processes would be completely out of control, inefficient, and certainly unprofitable. One of the most critical elements for an effective control system is the ability to measure the critical variables in the system in a time frame shorter than their natural inclination to change. This is a fundamental principle for the control of production processes, or any processes.

As has already been discussed, even human learning and performance are tied to these same control theory characteristics. Take the example of a person trying to catch a ball thrown to him or her. Since the ball continuously moves over its route to the person trying to catch it, it is essential that the person can almost continuously measure where the ball is, how fast it is moving, and over what trajectory it is travelling if they are to successfully catch it. Without these real-time measurement systems humans possess through their sensory systems (in this case sight) they would have little success in effectively completing the task. Just imagine trying to catch a ball thrown to you from 100 feet away if you can only open your eyes for a tenth of a second every minute. You might actually catch the ball every once in a while if you are lucky, but your success rate most likely would not be very great.

The point is that effectively managing an industrial business requires the application of control theory to the key business variables. Since most of the key industrial business variables did not change for months at a time over most of the last couple of centuries, the accounting compromise resulting in monthly measurement systems actually provided a fairly solid measurement basis for the effective management and control of industrial businesses. The compromise tended to work well. Business executives may have been more comfortable having more frequent visibility into the profitability of their business, but more frequent business measurements most likely would not have added much incremental value to the business. The business variables changed so slowly that managing and controlling the profitability of the business were not nearly as challenging as managing and controlling the production processes. As a result, most of the formal application of control theory was directed to the production processes, not the business.

A very significant organizational and operational consequence of the separation of the control of industrial processes (operations) and the management of business processes was that the organizations associated with managing the business and controlling the industrial processes evolved to be separate and quite independent from each other. The industrial operations were focused on effectively making the right decisions—either automatically or manually—in real time to drive the efficiency of the production operations. The business organization became very much associated with accounting for the business and producing the reports necessary to report those results to the executives, owners, and investors in a standard manner. This separation worked well. As long as the operations teams worked as efficiently as they could, the profitability from the operation would be maximized. The operations people were seldom, if ever, provided with information on the profitability of the business. This monthly information provided little guidance on how they might behave differently for the betterment of the business. From the plant floor perspective, the business information became available way too infrequently for them to even begin to understand how they might use it. As a result, operations management focused on efficiency and business management focused on reporting profitability. Once again, we are confronted with the unspoken assumption that efficiency equals profitability, and that maximum efficiency equals maximum profitability.

THE COMPUTER REVOLUTION

The computer revolution began in earnest in the 1950s, but truly started to impact industrial businesses in the 1960s. The first area of impact of computers on industrial companies was at the business management levels. As industrial firms grew and increased in complexity, many firms started to have multiple manufacturing sites and operations distributed over wide geographic areas. Each plant could be controlled and operated independently, but business results had to be reported corporately. The challenge of pulling all of the necessary information across all the sites together and producing the required reports was daunting and actually imposed practical constraints on the organization and growth of many enterprises.

The introduction of cost-effective digital computers was seen to be the solution and industrial companies started investing in the new technology almost as soon as it became economically feasible. Since the primary work focus of these new computer systems was directed toward business reporting functions, they were frequently referred to as Management Information Systems (MIS). The problem was that these computers were very complicated and required an entirely new knowledge base to make them work. Almost overnight new departments developed within industrial organizations to deal with the complexity and unfamiliarity of the computer. These new departments, often called the Information Technology (IT) departments, were typically comprised of programmers and system analysts who could develop the computer programs that instructed the computers on how to perform the desired tasks, operators who could actually run the computers, and maintenance personnel to keep the computers up and running. In most cases, the talent base in the IT departments did not understand the details of the businesses they were working to support, and for many, it was almost a badge of honor that they did not. One key challenge was for the business managers to explain exactly what they needed the computers to do in clear and concise language so that the programmers could correctly develop the desired programs. This has been an ongoing challenge in business and industry with many business professionals openly frustrated with the output of their own IT team.

Since the IT departments were almost exclusively focused on producing the necessary management reports for the business management of the company, they had little or no interaction with the actual plant operations. For the most part, they did not know or care what was happening in the manufacturing plants because it did not really impact how they performed their tasks.

IBM deliberately fostered this divide between IT and production by requiring its employees to wear suits and ties even on the production floor.

In the 1960s, the technology utilized for the control of the manufacturing and production processes was not based on digital computers. Rather, it was typically traditional electromechanical relays, gauges, analog controllers, pumps, and valves and had very little in common with the computers for which the IT departments were responsible. As a result, plant operations organizations and IT departments developed and evolved almost completely independently.

It did not take too long after the introduction of the computer for business management and reporting functions, for industrial engineering talent to look to the computer as an effective tool for the control and operation of the plants. Throughout the 1960s, 1970s, and 1980s, digital computer-based manufacturing and process control systems, such as programmable logic controllers (PLCs) and distributed control systems (DCSs), were developed by the instrumentation companies and electrical suppliers and came into common usage on the plant floor. These control systems were designed with software operating systems that allowed them to respond to changes in plant floor variables in real time frames as compared to the business reporting-based operating systems of the MIS computers. The plant floor systems were commonly referred to as real-time systems, while the MIS systems were called transactional systems due to the reporting-based transactional design of their operating systems.

Nonetheless, in the matter of a very few years, the technology base of the management reporting functions of industrial businesses and the plant floor controls started to converge. This technological convergence was seen as a huge opportunity by the technologists in these companies. They thought that the information in the real-time systems might help make the tasks being done in the MIS systems easier and better. They did not seem to understand how this could be done, but they were convinced that connecting the plant floor systems to the business systems would accomplish something good. This was a classic technology push mindset. That is, the technology allows us to do it, so we should go ahead and do it and see if there is any business value after it is done. There are numerous examples of industrial initiatives over the past few decades that exemplify this mindset, including the current push toward mobile computing and wireless sensor technology.

Perhaps, the most amusing was a movement in the 1970s and 1980s called Computer Integrated Manufacturing (CIM). The fervor around this movement rapidly reached a boiling point. It was difficult in the early 1980s to attend any industrial conference without 25% of the papers and talks focused on CIM. In fact, during this time, I (Peter) was actually asked by my company's leadership team to be the vice president of CIM. Of course, I agreed, even though I was not quite sure what CIM was or why it might be good. You just cannot turn down a vice president position! So I became vice president of something and I was not quite sure what it was that I was vice president of. As I matured in my business career I learned that this happened more frequently than I would have supposed. After a couple of years of trying to determine what CIM was I finally figured it out. CIM meant: "if you connect all the computers in an industrial company together—something *good* is bound to happen." Nobody seemed to quite know what *good* might happen and I was never convinced that it did much good. However, a considerable amount of money was spent CIMing (using the verb form) and there seem to have been a lot of CIM experts—including me. Once I learned what CIM was my job changed and I was no longer vice president of CIM.

The reason I decided to call out this example of the technology-based industrial initiative called CIM is that it is not unique. Over the past 30 years after the introduction of the computer to industrial operations, computer-based technologists, both in the IT and engineering departments, have been driving a number of initiatives around specific computer technologies that no one seems quite sure will actually help either the operations or the business. Other examples include expert systems, SPC (Statistical Process Control), MES (Manufacturing Execution Systems), cloud computing, lights-out manufacturing, and wireless technology. It is not that these technology-driven initiatives did not offer some potential value—they did; it is just that the technologists often seem more interested in pushing the technology rather than in solving industrial operations or business problems. Although to a large extent this technology focus continues today, there seems to have been a transition to more problem-solving content to many discussions rather than just a pure technology focus.

In spite of the focus associated with connecting the plant floor to the business systems, for the most part, the plant floor operations and the business systems operated quite independently—and still do. There are good reasons for this. IT organizations are typically well versed in computer system technology, but not the business and not the plant floor operations. They take their directives from the business leadership and try to set up and operate the MIS environment to meet the business reporting needs. Plant floor engineers, on the other hand, understand the plant floor control systems to the level of being able to apply them to engineering challenges in the plants. As a result, the three organizations—business management, IT, and plant engineering—speak different technical languages, know different things, and have great difficulty communicating and cooperating since they work on different time bases. This has not been too much of a problem historically since they could operate somewhat independently and still succeed to a reasonable degree, but industrial businesses have recently undergone a drastic, although visibly subtle, change requiring these three organizations to cooperate as never before.

BUSINESS—OUT OF CONTROL

Since about the mid-1990s, a subtle but very important change has impacted the business of industry throughout the world. Perhaps, the trigger mechanism of this change was the deregulation of electrical power resulting in the commercial opening of the electric power grids. This opening of the power grids happened in different geographies at different times and in some countries is still taking place. There was considerable news coverage as legislation associated with the deregulation of the electric power market was passed by different government bodies. The news reports often projected positive optimism about the opportunities this afforded to small independent power producers who would be able to sell their excess electricity to the power grids and, thereby produce additional income. It was not unusual for private citizens to purchase power generation sources, such as windmills or solar panels, with the hope of reducing their electricity costs and making additional income. All of this was very good and actually one of the intentions of the legislation. With deregulation comes increased competition. Increased competition leads to greater commercial fluctuation. Electricity producers, whether private citizens or large corporations, started to realize that they could maximize their profit from their electricity production by charging their electricity to the grid during high-price periods and reducing production during low-price periods. Pricing volatility in the electricity market increased significantly. Electricity consumers, who had been used to relatively fixed electricity cost during the time of regulated production, started to feel the impact of the rapid price fluctuations and applied pressure on governments to help. Many governments passed regulations limiting the frequency of the fluctuations. In the United States, for example, the electricity price fluctuations were limited to every 15 minutes in order to provide some level of stability.

The deregulation of the electricity market resulted in a domino effect across many other markets. Since the production processes for other types of energy (natural gas etc.) are highly dependent on electricity, they started experiencing frequent fluctuations as well. In the United States, the price of natural gas can change on a 15-minute basis, very similar to electricity. Many commodities, such as base metals that use both electricity and natural gas in their production and are also sold on deregulated basis are experiencing price fluctuation time frames of under a minute. These frequent fluctuations also propagate through industries that utilize these energy sources and commodities in their production processes. Today, many of the *business* variables of industrial operations are fluctuating at rates almost as frequent as the real-time rates in the production operations—real time. Over the past decade, the business of industry has transitioned from highly transactional dynamics to real-time dynamics.

As the transition toward real-time industrial business started to impact industrial operations, it appears as though most industrial companies paid little or no attention to it. There could be many reasons for this. Perhaps, the difficult market environments and economic downturns that were simultaneously impacting industrial businesses were so extreme that they provided a smoke screen over the transition to real-time business variables and, as a result, business executives did not notice the transition. Perhaps, executives recognized that they did not appear to have as much control of their businesses, but noticed that other companies seemed to be experiencing similar conditions and they just assumed some temporary market condition was at work that was beyond their control. The plant floor operations personnel who might have had the best view and understanding of what was happening were blind to it because they were focused on efficiency and not on the business variables. Perhaps, IT teams had been more focused on the technology that the business was deploying or investigating, as was their job, and did not notice the business transition. In any case, as the business of industry steadily transitioned to real time, it did not seem to have been noticed and very little was done to respond.

A LEAD BUSINESS INDICATOR—THE ELECTRIC POWER INDUSTRY

Since the electric power industry was the first domino in the overall domino effect leading to the transition of industrial businesses to real-time variation, it may be worth taking a closer look at how this industry has responded. The response of the electric power industry may serve as an example for other industries as they transition to real-time business variability. For decades, electric power production was accomplished by large power generating plants designed to operate continuously to produce as much electricity as possible to meet market demand. Since the price paid for electricity was highly regulated, there was little need for agility across the distribution network from generator to consumers. Typically, the electricity generators controlled the generation plants and the transmission and distribution facilities across a defined geography. The power generation, transmission, and distribution systems were fairly stable for long periods of time. Power companies would make money according to the regulators formulas. Although the electric power systems worked well, attaining higher levels of efficiency to gain higher profitability was not typically the chief concern, because in a regulated system, higher efficiency did not directly translate to profitability.

With deregulation the dynamics across the power generation, transmission, and distribution networks changed dramatically. The power distribution systems had to open up to be able to receive generated power from a variety of sources beyond just the large power generators—including both industrial and private generation sources. This led to logistic issues for the traditional power companies, especially with respect to the transmission systems. In many areas, the management of the transmission systems was assumed by separate power grid management companies. These grid management companies specialized in the management of the electricity supply, electricity demand, and pricing levels across the transmission network. Grid management became a highly specialized technical and management business model in its own right.

Traditional power generation companies typically made up of a number of power generation stations originally designed for almost steady state, continuous electricity production found themselves requiring news level of agility. In order to maximize their profitability, they had to maximize their output during periods of high-pricing levels on the grid and minimizing their output during low-price periods. Traditional power stations were not designed to accomplish this. With deregulation of the power industry these companies found that their profitability was going out of control even if their plant efficiency was still in control. But the power companies realized that even with considerable variability in supply and demand across the grids, which actually drove the pricing fluctuations, there would be a need for a base load of power at all times. Tearing down traditional power stations and replacing them with more agile stations was not commercially feasible. Also, the traditional power stations, operating in a continuous and stable manner, were found to produce fairly economical power. The power companies realized that they could provide a base load of power to the grids economically through the effective utilization of the traditional power generation stations and attain the necessary levels of agility to manage peak loads by building new, more agile, and typically smaller power generation stations, often gas powered combined cycle stations. The smaller, "peaking stations" could be started up and shut down fairly quickly, but they tended to be more expensive to operate.

Operating the peaking plants only made good economic sense when the pricing levels on the grid increased to the levels at which the generators could operate them profitably. This resulted in an interesting commercial sort of dance between the power generators and the grid managers. As the grid managers determined that demand would increase requiring a higher level of supply, they had to increase the price of electricity on the grid to enable the power generators to operate their peaking plants profitably.

The pricing dance did not only affect the electricity producers and grid managers, but certainly also the electricity consumers. The price consumers had been paying for electricity in the regulated environment had been stable for months at a time. With deregulation the price started to vary across a single day. In the immediate period following deregulation consumers found that they could easily predict the high-price periods by predicting the high-electricity-usage periods. Many started to realize that the price would go up early in the morning as people were waking up and using increased energy for showers and breakfast. The price would also go up during dinner time as lights were turned on, and evening meals prepared. But then consumers started to adjust their behaviors to capitalize on the lower price periods of the day and avoid the higher price periods. The real-time supply and pricing dynamics on the grid were further exacerbated by all of the other energy producers supplying the grid who were trying to dance the same dance as traditional power generators. As a result, the supply and resulting pricing profile on the grid adjusted and became more variable and much less predictable. Many consumers, including industrial consumers, are not aware of this increased variability since they do not have a view of the real-time electricity pricing changes, and still believe that avoiding the traditional high-price periods of the day is the key to minimizing what they pay for their electricity. Unfortunately, this is not necessarily true, and the lack of real-time electricity price response is leading to much higher electric bills for these consumers than may be necessary.

Today the electric power industry may represent the most real-time business in the industrial world. Although there are still many kinks that need to be worked out in the electricity supply, transmission, distribution, and consumption value chain, it serves as a good lead indicator of what industrial companies need to consider as they move forward to effectively respond to the real-time business transition. Certainly, the business dynamics in industrial companies in other segments are quite different from those of the electric power industry, but one thing is clear—all industrial businesses are moving more toward real-time variability and must respond to this transition to regain, maintain, and sustain their profitability.

REAL-TIME INDUSTRIAL BUSINESSES

When analyzing the real-time business transition across all industrial segments, the example of the electric power industry is particularly poignant because not only was this transition triggered in this segment, but this segment behaves, to a large extent, like almost all industrial businesses. As the impact of real-time business is analyzed across industry in general, the electric power industry plays the unique role of both primary cause and resulting effect. From this point forward, therefore, for the sake of clarity, the electric power industry will be viewed and discussed from the perspective of primary cause. We hope that this helps drive clarity without loss of context.

Most industrial plants have a very general production model that defines how they operate as shown in Figure 4.1. Although this model presents a gross simplification, it also generally holds true and is useful for understanding the

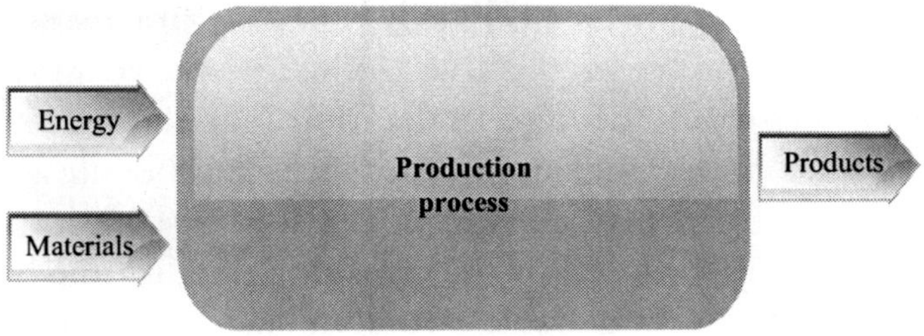

Figure 4.1. Simplified general production model. All manufacturing and production processes can be thought of as transformations of energy and materials into productions.

impact of a general market driver, such as the transition to real-time business. Manufacturing and production plants are designed to make products by precisely combining raw materials and energy through a structured production process. This is generally true for all production types and industries. Different industries are often described according to the characteristics of the basic production processes required to produce the products. The continuous process industries, such as oil refining, petrochemicals, power, and paper, are predominantly comprised of continuous production processes, utilizing primarily fluids as raw materials and consume relatively large quantities of energy in the production. The batch process industries, such as specialty chemicals, pharmaceuticals, processed foods, and beverages, are predominantly comprised of independent but connected vessels and units, produce the products in batches or lots, utilize both fluid and solid raw materials, and consume relatively less energy than their continuous counterparts. The discrete manufacturing industries, such as automotive, aerospace, and heavy equipment, are predominantly comprised of assembly lines that assemble numerous component parts into products and consume relatively little energy.

The general impact of the real-time business transition on industrial operations was felt first in the energy intensive continuous process industries, as might be expected, since the volatility of the energy market led to the cause-and-effect chain driving the transition. Initially, the impact was not noticed in most continuous production plants for a number of reasons. First, the impact of energy price volatility was felt at plant levels of continuous production operations and, as has been previously discussed, there was no visibility to the business variables directly associated with profitability among the plant floor personnel. The employees in the best position to see and deal with the volatility had no visibility of it. They had traditionally focused on increasing the efficiency of the operation, not the profitability. Second, the initial price variations triggered by the deregulation of the power industry, although fairly frequent, were also quite small. The initial impact on the profit of the operations was likewise quite small. Finally, the daily pricing volatility in the electricity market was pretty predictable. Industrial businesses learned early on that energy prices would be high in the mornings and evenings as households increased their energy consumption and would schedule as many energy intensive operations to take place during the low-cost periods. Therefore, either the initial problems caused by the transition to real-time business variation were so small or so predictable that they had minimal impact on the profitability of industrial operations, and the impact they did have was easily manageable.

Often, when a problem has been initially identified and theoretically effectively addressed, as was the initial case with the transition to real-time business variability, people consider it solved, and tend to ignore it from that point forward. To a large extent, this is exactly what happened with these industrial operations. Unfortunately, the problem did not stay the same. With time, the dynamics associated with real-time business variability changed drastically and industrial companies tended to ignore them. As a large number of electricity consumers identified the initial price repeatability across a day with higher electricity prices in the mornings and evenings, energy consumption models changed with higher levels of consumption during the "low-cost" periods. Over time, not only was the repeatability lost, but the electricity pricing across any time period became very difficult to predict. With increasing competition in the electricity market, it continued to increase in volatility driving much greater pricing changes than had been initially experienced. Finally, since the production of many energy and raw material sources depend on electricity in their production processes, the price of other energy types and some raw material sources started to exhibit the same type of real-time volatility. Combining multiple energy and raw material sources together in production processes, each experiencing its own real-time pricing volatility, created a real-time profitability mess for industrial companies. Since many believed that they had already addressed the real-time business issues resulting from electricity deregulation by scheduling energy intensive activities to take place in the initial low-cost periods, they did not look to that as a contributing factor. Many industrial executives knew that their profitability was out of control, but had no idea on what to do about it!

THE IMPACT

As industrial business variables started to experience real-time variation, the variation began impacting the profitability of the industrial organizations. Many leading industrial companies had developed engineering and operations management teams over decades that had become absolute experts at controlling plant efficiency. The concepts behind controlling efficiency were fairly straightforward even if the technologies and techniques necessary to do it well were very sophisticated. Controlling efficiency essentially involves making the desired quantity of products at the desired quality at as low a cost as possible. Minimizing costs involved cutting head count, materials consumed, and energy consumed among other smaller impact items. Hence, the engineering departments of the energy intensive industries often worked on energy management programs designed to reduce the overall energy consumption and thereby directly reduce the cost of energy.

Energy management programs of this type worked quite well in the era of business variable stability that began with the start of the Industrial Revolution and continued until just a decade or so ago. During this period, reducing the energy consumed per production unit produced directly translated into a resulting profitability improvement. The plant floor personnel did not even need to know what the cost of energy or the actual profitability was. They just needed to reduce energy consumption and—voila—profitability improved.

As real-time energy pricing volatility started to impact industrial operations, the traditional focus on reducing overall consumption did not appear to be working—at least from the perspective of improving profitability. For the first time in history, energy consumption could be reduced while the overall cost of energy actually increased. This is due to the fact that the price the plant was paying for the electricity was highly variable over very short time frames. Suppose a plant operation consumed less energy over a day, but the bulk of the energy they did consume was consumed during the high-price periods of the day, their cost might actually increase. This became visible to business managers who actually had a view of profitability much earlier than it did to the plant operations teams. Business management started to exert pressure on the plant teams to help reduce profitability as well as consumption. This was driving engineering and operations management teams out of their minds. The things they had done for decades and that they were so good at executing did not seem to be working any more. It was not that energy was just continuously increasing. Rather, there did not seem to be any relationship between the energy management programs and profitability. Profitability appeared to be varying all over the place.

The problem was that the price industrial operations were paying for energy was changing multiple times an hour and the plant personnel had no visibility of the price fluctuations. In fact, in most industrial information and automation systems the price of energy had been considered to be within the domain of the business systems, which had typically been developed for monthly reporting. Plant engineering teams had a conundrum on their hands. They were excellent at real-time control and very adept in applying control theory to plant efficiency by controlling the physical, biological, and chemical variables in the plants. This was exactly the skill set that needed to be applied to the control of profitability, but they did not have access to the measurement information in the time frame or space resolution necessary to control the profitability. A fundamental principle of control theory is that to effectively control any variable it is required that the variable be measured in a time frame that enables control. This just was not the case and there was no clear path available to get the necessary data. The net result was that, although the plant efficiency was in control, plant profitability was out of control!

A simple example may be useful in clarifying this situation. There was a fairly profitable zinc operation that I (Peter) had the privilege to work with at one point a number of years ago. It was a very well-run plant with very good efficiency, but their profitability was all over the place. The plant had an ore availability problem as the local ore body was not being mined anymore and they bought ore and had it shipped in from a number of different suppliers. As a result, the flow of ore to the plant was inconsistent causing them to have to shut down the zinc roaster a few times each month to wait for ore. Zinc roasters are huge energy consumers. In fact, the roaster at this site tended to consume about 40% of the output of the local power plant when operating normally. During startup, the roaster would consume even more electricity than it did while normally operating. The plant operations people had no insight into the pricing levels of the electricity supplied from the local power plant and would just start up the roaster when needed. The cost of electricity to the plant was going up dramatically.

Talking to the management of the local power plant we learned that the starting up and shutting down of the roaster was making it almost impossible for them to manage the supply on the transmission system. They had no idea when the zinc plant was going to start up and shut down and could not effectively adjust for it. When the roaster suddenly came offline, they found that they were producing way too much electricity for the needs of the community. When the roaster suddenly started up, they had to get their production output up very quickly—which was a huge task. The management of the power company indicated that the result of this was that they were not as efficient or profitable as they needed to be. We asked if they could provide a pricing schedule for each day prior to the start of the day so that the zinc plant could plan its startups during low-price periods. They said they would and guaranteed the pricing if the zinc plant would target low-cost periods. The cost cycles for electricity were set up to match the predicted demand cycles. Therefore, the low-cost periods were the low-demand periods that were the best times for the power plant to have the zinc roaster come online. The deal was struck. The result was that the zinc operation found they could often start up the roaster for over $20,000 less in electricity cost each time. The power company found that they could manage the demand loads much more effectively. All principles in this rather compact value chain benefited—including all the other energy consumers service by the power plant.

This is a fairly simple illustration that actually occurred in the early stages after electricity deregulation. However, the basic principles hold true for industry in general. The business variables of industrial operations are experiencing considerable variability in real time. This increased variability is driving profitability out of control. The solution is not to ignore the problem, rather to face it head on.

REAL-TIME PROFITABILITY MODEL

Anyone who has reviewed an annual report of any company realizes that there are many variables that are associated with the performance and profitability of industrial companies. If all of the variables presented in a standard balance sheet or profit and loss statement varied on a subminute basis it would be overwhelming to the business to provide effective control. Clearly, not all variables associated with the profitability of an industrial operation are experiencing highly frequent levels of fluctuation. It is important to frame the issues correctly and take a step back and understand what variables associated with profitability actually experience variability in a subday time frame.

A close analysis of industrial profitability reveals that there are only three critical variables directly associated with profitability that fluctuate on a reasonably frequent basis: production value, energy cost, and material cost. To be clear, these are not the only variables associated with the profitability of the operation. For example, labor cost is a critical component of profitability, but unless the organization has a horrendous turnover problem, labor costs should not vary all that much within monthly time frames and actually should not even vary much month-to-month. Production value is the actual market value of the products as they are produced. This may not seem to be a variable that would experience real-time fluctuation, but a short monitoring of a commodity price ticker on a financial network will clearly show that the market price of many industrial commodities, such as metals, and gas change multiple times per hour. The frequent fluctuations of energy costs have been previously addressed. Since many of the materials, such as metals and petrochemicals used in industrial production, come from other energy-intensive production operations, the price paid for these materials tends to fluctuate at a fairly frequent rate. Optimizing the real-time components of profitability involves maximizing the production value of the operations while minimizing the energy cost and material cost. This relationship is referred to as the real-time profitability model for industry (Figure 4.2).

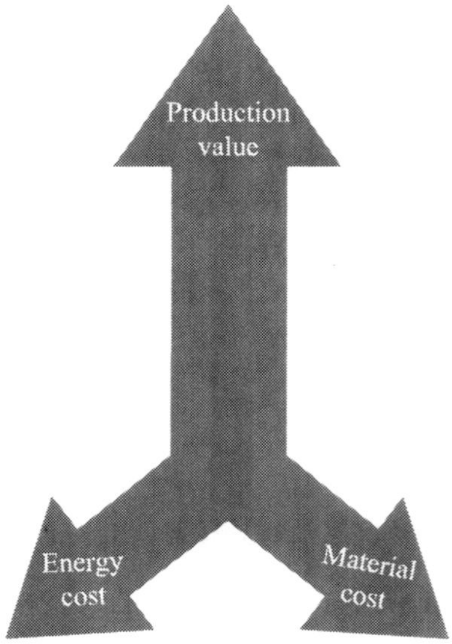

Figure 4.2. Industrial real-time profitability model. The three components of industrial profitability that demonstrate real-time variability are production value, energy costs, and material costs. The primary business objective is maximizing the production value, but production value must be balanced with energy and material costs to maximize profitability.

The industrial real-time profitability model represents a multiple objective optimization problem. The three objective functions are (1) maximizing production value, (2) minimizing energy costs, and (3) minimizing material costs. Traditional industrial optimization approaches have been limited to single objective optimization problems because the mathematics required to execute multiple objective problems is so complex. Multiple objective problems have been represented as single objective by converting all but one of the objective functions into constraint functions. This is very limiting in highly dynamic industrial environments in which all three objectives continually interact with each other. Effectively controlling the real-time profitability vectors must involve balancing all three objectives for maximum profitability in a continuous manner.

As with any optimization problem, even a three-objective problem would be fairly easy to solve if there were no constraints on any of the vectors. In this case, the solution would be to produce as many products as the market can absorb at all times while reducing both energy and material costs to zero. Obviously, this cannot be practically done because of the constraints on the system. The key to dealing with an optimization problem of this type is to identify the constraints that are most limiting on the solution set and optimize within those constraints. In the case of the real-time profitability model for heavy industrial environments, the most limiting constraints are typically limitations of physical equipment, safety, and environmental integrity. Since well-designed plants typically encounter practical safety constraints—whether safety of plant, people, or environment—prior to physical constraints, safety can commonly be considered to be the primary constraint function to real-time profitability for industrial environments (Figure 4.3). Interestingly, the safety risk for an industrial operation may change over time, and in different operating modes. Therefore, even the constraint envelope may be variable and may experience real-time fluctuations. This makes the optimization problem even more challenging.

Since the constrained real-time profitability model is comprised of at least four competing components (production value, energy cost, material cost, and safety) each of which experiences real-time variability, the first problem that must be resolved to be able to effectively address profitability is one of time. All of these basic variables must be measured in a time frame associated with their potential change frequency to have any chance of effectively managing the real-time profitability components. The second problem is one of space. Today, industrial operations tend to resolve the profitability or their

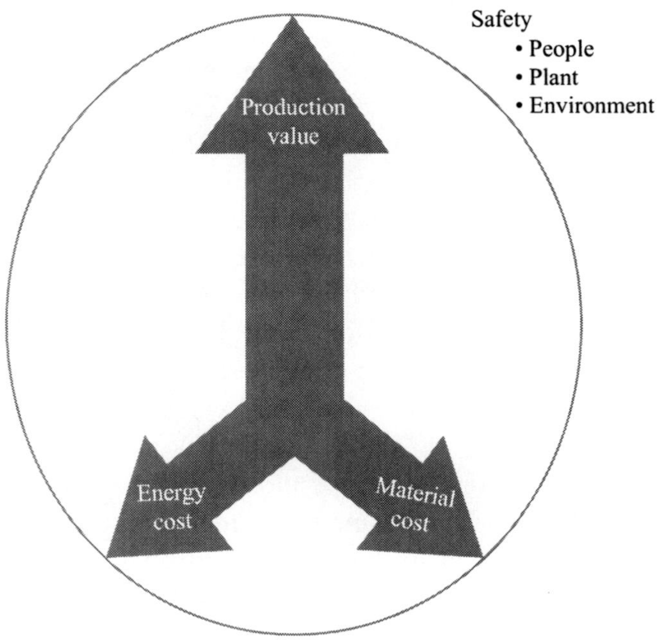

Figure 4.3. Constrained industrial real-time profitability model. In the heavy process industries the primary constraints on real-time profitability are safety of people, equipment and environment. These constraints are typically encountered before equipment or process constraints.

manufacturing and production operations across an entire plant or factory. Most people dealing with the real-time operation of the plants have responsibility of a much smaller scope. Even if the critical measures of the components of real-time profitability were provided on the right time frame, providing plant-wide data to operators, maintenance personnel, or engineers responsible for a small subset of the overall operation would not enable effective decision support. The reason is that multiple actions by multiple people may be taking place across the plant at the same time. It would be near impossible for any operator to discern if their actions had the desired effect. What is necessary is the right information provided on the right time frame and contextualized to the operating domain for which each person is responsible.

TRADITIONAL ORGANIZATIONAL AND ARCHITECTURE MISMATCH

Effectively meeting the challenges resulting from the shift toward real-time profitability will require fundamental changes in both the technology organizations and architectural approaches in industrial companies. The traditional assumption upon which both technology organizations and architecture were built was that the business functions of industrial organizations could be resolved monthly at business management levels and operational functions could be resolved in real time at the plant levels. This assumption was based on the idea that resolving business functions at management levels on a monthly basis would be adequate for both effective business decision making and financial reporting functions. In fact, for the most part, this assumption was true when the business variables required for both business decision making and for financial reporting were stable for monthly or greater periods of time. When the business variables did not effectively change within monthly periods, monthly business information provided for reasonable business decision support.

Traditional technology organizations and architectures were both designed to support the assumption. The business management systems were managed by IT organizations and the plant floor systems by engineering organizations. Very little business information made its way into the automation systems and in spite of the focus on plant floor to Enterprise Resource Planning (ERP) integration of the past 20 years, very little plant floor operational information made its way into the business management systems. Both the technology organizations and architectures were typically perceived as horizontal layers (Figure 4.4) with the operations management functions picture below the business management structures.

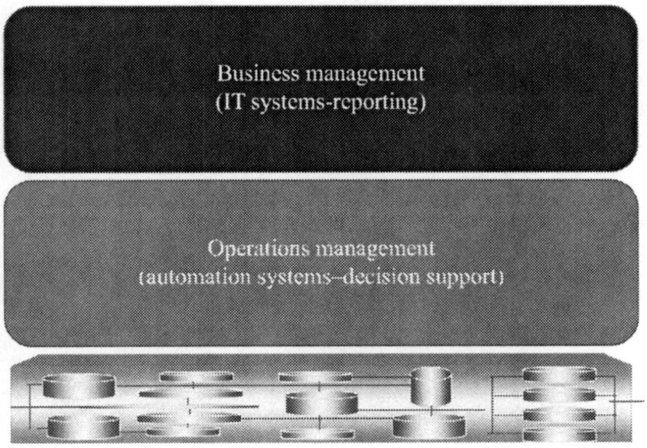

Figure 4.4. Traditional industrial information management. Most industrial operations perceived two levels of information management; business information management, which is the responsibility of the Information Technology teams, and operations management, which is typically the responsibility of the engineering department.

As time progressed and more industrial application software was developed to meet more industrial functionality, the number of horizontal layers was often increased. One common representation included a software layer in between the business management systems on the top, and the automation systems on the bottom, to depict functionality that neither system seemed to include (Figure 4.5). This middle layer has been commonly referred to as Manufacturing Execution Systems (MES) since this phrase was suggested by Advanced Manufacturing Research (AMR) in the mid-1980s. Interestingly, even as this three-layer model became a more accepted depiction of industrial automation and information architectures, few industrial organizations actually reorganized to the model. Instead, the IT organizations and the plant engineering organizations often found themselves fighting over the middle MES layer with some of the functions going to the IT organization and some to the engineering organizations. Many a times, the same functional software was acquired multiple times by IT and by engineering. Software and system vendors also fought over the MES space. ERP software vendors who traditionally provided the business management functions and industrial automation system and software vendors both started acquiring smaller software companies offering MES functionality. Confusion reigned both in industrial companies and in the supplier community and this confusion continues today. This type of architectural and organizational battles is often indicative of a much larger trend. This was certainly the case with the industrial market.

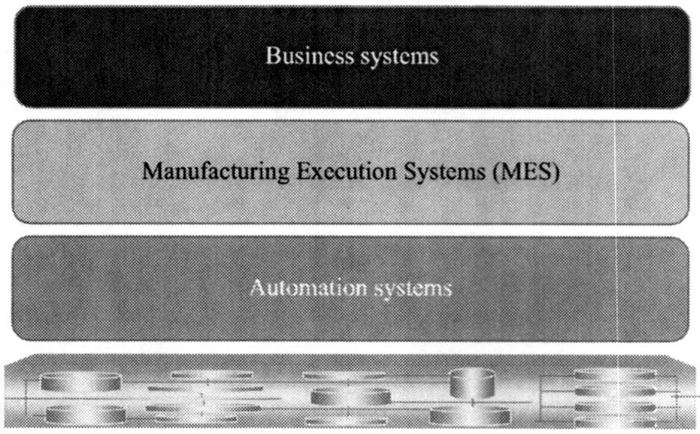

Figure 4.5. Three-layer industrial automation/information model. A number of automation/information functions and applications were developed that were not traditionally part of the business or automation systems. This led to a third industrial information layer commonly referred to as the Manufacturing Execution System (MES) layer.

Much of the discussion with respect to the appropriate architectural and organizational placement for software offering some aspect of MES functionality was whether the function would provide more value if run in real time, or whether it was more appropriate to run it periodically on human schedules (transactional). If the function required real-time response it would often be placed within the engineering/automation system domain. If it was better run on a fixed periodic basis, it would typically become part of the IT/business system domain. For many MES functions, this was difficult for industrial companies to determine.

As has been previously discussed, over the past 10–15 years, many of the business variables used to control the profitability of industrial organizations have transitioned from transactional to real-time variability. As this transition took place the software functionality had to correspondingly transition to real-time operations if the new real-time business variables were either inputs to, or outputs from, the software. That is, as the transition toward real-time business variability took place, more and more of the MES software modules that had been effective operating on a transactional basis needed to operate in real time to be effective. This was one of the background driving forces that was causing so much confusion at the MES levels.

It is helpful in trying to determine the correct operational environment for any industrial function being supported by software to understand the underlying informational requirements the software supports. In a slight oversimplification the informational requirements of industrial operations can be partitioned into two basic kinds of information: reporting and decision support. Reporting information is required to provide records of what has happened in the industrial organization to both internal and external constituencies for analysis. Decision support information is required to make sound decisions for effective operation and profitability. There is certainly some overlap between these two categories of information, but as the business of industries has moved to more real-time variability, there has been a requirement shift toward more real-time decision support information from the plant floor all the way through the business management of industrial enterprises. In fact, it does not require much analysis to realize that there are both decision support and reporting requirements at all levels of industrial organizations. Some examples of plant floor reporting requirements are environmental regulatory reports, plant configurations data, and safety reporting. With this in mind, perhaps, a more reasonable architectural partitioning than the traditional horizontal approach would be a vertical partitioning of reporting information and decision support information (Figure 4.6).

The traditional mindset that led to the horizontal partitioning of industrial organizations has actually become a barrier for industrial organizations moving forward. If all of the real-time functions are relegated to the automation systems, but no business functions are performed in the automation systems, where will the emerging set of real-time business functionality be performed? Conversely, there are many business functions that directly impact the operational performance but are currently reserved for the IT systems. Both the installed system architectures and the prevailing mindset in industrial organizations that led to the mindset are presenting huge barriers, as industrial organizations need to adjust to meet the challenges brought on by the transition to real-time industrial business variability.

Not only do the technology architectures present a barrier to meeting this challenge, but the organizational constructs associated with the traditional separation between automation and IT might even provide greater obstacles to

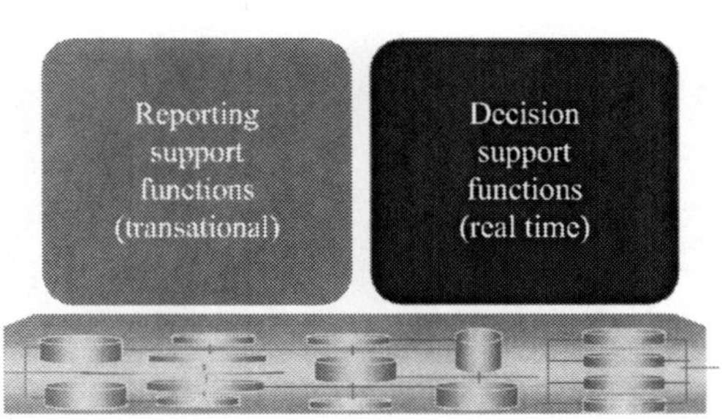

Figure 4.6. Performance-based industrial information management. The traditional horizontal layering of the industrial automation/information model is insufficient for performance-based industrial perspectives. A vertical perspective emphasizing that both real-time and transactional functions is required.

success. There are major differences between the talent that is responsible for the IT systems and the talent responsible for the automation systems. IT talent is well versed in technology architecture, infrastructure, and application deployment. Engineering talent is well versed in meeting plant challenges. IT excels in transactional computing while engineering excels in real-time computing. Both of these skill sets are required for success in meeting the challenges of the emerging real-time industrial businesses. Unfortunately, over the decades, many IT and engineering organizations have developed a distrust of each other. Both IT and engineering seem to believe that the other is encroaching on their territory. In recent years, a number of industrial organizations have begun to merge the IT and engineering organizations into a single technology team. Although there are many challenges associated with this, those who have done this well are finding that both the IT and engineering talent sets are required to meet the variety of challenges throughout the business and across industrial enterprises. Those industrial companies that have merged these two functions well are finding themselves at an advantage as industrial businesses move more and more to real-time variability.

Over the past decade the business of industry has moved toward real-time variability, which has caused profitability to be out of control even when plant floor efficiency is well controlled. Trying to meet this challenge by continuing with the same architectural and organizational approaches that have worked in the past will not do. Major changes are required both in terms of technology and organizational design. Industrial profitability is out of control. Those industrial companies that survive will be the ones that understand this issue and rise to meet it with new technological and organizational approaches.

5

Applying Control Theory to Profitability

Humans utilize feedback control in almost every aspect of life. In fact, human learning at its basic level is based on feedback control. We often discuss the basics of learning as "trial and error." Trial and error is a feedback control system. Therefore, in almost every area in which learning takes place—reading, writing, arithmetic, riding a bicycle, hitting a baseball, and brushing our teeth—feedback control is the mechanism we utilize.

The efficiency of industrial operations has been controlled for many decades using feedback control systems. The concepts behind feedback control are really quite simple. First, a structured and repeatable process that needs to be controlled is required. For example, a child trying to use a baseball bat to hit a thrown baseball is a repeatable process if the ball is tossed over and over again; so the child might learn how to hit it with the bat. In an industrial operation a process that may need to be controlled might be the flow of a material through a pipe or the level of a liquid in a tank that has continuous inlet and outlet. The concepts associated with controlling the process of hitting a ball with a bat and controlling the value of an industrial variable are actually quite similar.

Second, whatever it is that is to be controlled must be measured both accurately and in a time frame that allows appropriate and effective action to take place. For example, if a child is attempting to learn how to hit a baseball with a baseball bat, the trajectory and speed of the ball must be measured by the child on a time frame that enables the child to effectively swing the bat to hit the ball. This is accomplished through the child's vision. If the child can see the ball continuously as it travels from the pitcher to home plate, in real time, with time we might expect that the child may learn how to hit the ball with the bat. On the other hand, if the child can view the ball in a blink of the eyes once every 10 seconds, the probability that the child will successfully hit the ball is significantly reduced. The point is that it is not just necessary to measure the trajectory and speed of the ball, but it is necessary to measure the trajectory and speed essentially continuously in real time. In the example of an industrial process the measurement system is likely to be a sensor that senses the aspect of the process to be controlled, such as flow, level, temperature, or pressure.

Third, once a continuous real-time measurement system is in place, a control mechanism that uses the measurement system must be established. In the case of the child trying to hit the baseball, the control mechanism is a complex combination of the brain, the nervous system, the muscles, and the skeletal system. The child learns to swing the bat at the ball and the vision conveys how successful the operation was by how close the bat was to hitting the ball. The child's brain adjusts the swing response in an attempt to get closer. The child repeats this process of trial and error, hopefully swinging closer to the ball with each attempt, until the child successfully hits the ball.

In early industrial systems the controller was typically a human operator who viewed a gauge that conveyed the value of the process variable of interest. The operator would adjust a valve to attempt to make the value of the process variable equal to the desired value. This process was very similar to the process of a child trying to hit the baseball. Over time, automatic controllers were substituted for human operators, but performed essentially the same function. A transmitter associated with the process sensor would transmit a signal representing the value of the process variable to the controller. The controller would compare the actual value to a predetermined desired value, the set point, and, if necessary, transmit an output signal to adjust some controlling mechanism, such as a valve, to attempt to alter the value of the process variable to equal the set point. Typically, the controller would initially adjust too much and the value of the process variable might overshoot the set point. In fact, the amount of adjustments the controller would make would be tuned so that the process would adjust as quickly as reasonable while the overshoot would be minimal. Since the controller would continue to adjust the valve, eventually, the value of the process variable would be made to equal the set point.

The fourth component of a feedback control system is a controlling mechanism. In the case of the child attempting to hit a baseball with the bat, the controlling mechanism is the swing of the bat. If it occurs at the right time in the right location the bat will connect with the ball. In the case of an industrial process, the controlling mechanism may be a valve, a variable speed pump, an adjustable drive, vents, or any number of other controlling mechanisms. The signal from the controller is transmitted to the controlling mechanism, which is adjusted to alter the value of the measured variable. The combination of the four components in an industrial feedback control system, the process, the measurement, the controller, and the controlling mechanism, are together referred to as a feedback control loop (Figure 5.1).

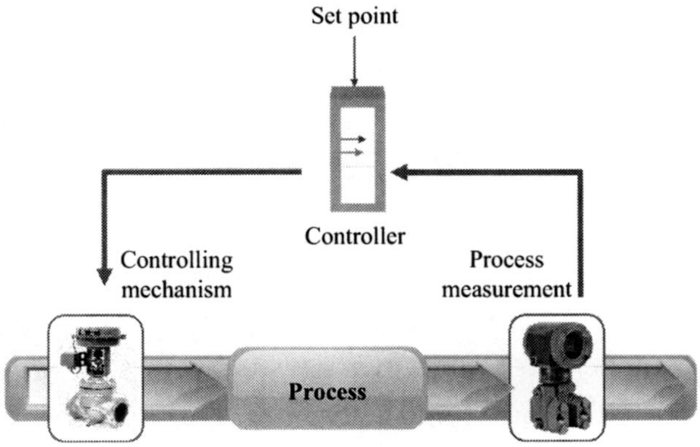

Figure 5.1. Process control loop. The basic process control loop is comprised of four fundamental components; a process, a measured variable, a controller, and a controlled variable manipulated via a controlling mechanism such as a valve.

CONTROLLING PROFITABILITY

In many respects, business processes operate similarly to industrial operational processes. They are comprised of a number of control loops, each of which needs to be controlled to a desired value, or set point. If each is controlled to the set point, the objective of the loop—which is typically to maximize profitability—can be effectively realized.

Most industrial professionals do not seem to think of maximizing profitability as a control problem. Rather, for decades, industrial companies treated maximizing profitability as a management problem. Sudipta Bhattacharya, who was then a senior vice president of SAP the German ERP software giant, once asked me (Peter) if I knew the difference between control and management. I had not previously thought about it and told him I guess I did not. He said, "Management is when you cannot control something." Although this appears to be a simple and obvious response, upon reflection it is really quite profound. Traditionally, all of the key economic variables comprising the profitability of industrial operations were stable over extended durations. Process variables changed frequently and control theory was applied to the process variables to maximize the efficiency of the operation. Since the economic variables were unchanging, controlling the efficiency of the industrial process directly translated into controlling profitability. In other words, managing the profitability of the operations was tantamount to controlling the profitability due to the dynamic characteristics of the operation. When the frequency of variation of the economic variables started to increase, although operational efficiency might have been in control, profitability was out of control. Industrial managers tried to *manage* it because they did not know how to *control* it. However, in most cases, they failed. In this case, they were actually managing profitability because they could not, or better yet—did not know how to control profitability.

As profitability transitioned to become more and more out of control, industrial managers were at a loss as to how to respond. The first aspect of solving any problem is to understand the kind of problem that is to be solved. Maximizing the profitability of industrial operations in a real-time business environment is a control problem. Therefore, control theory must be effectively applied to this problem to solve it.

In order to discern how feedback control can be effectively applied to industrial profitability, it is essential to identify the components of industrial profitability that are fluctuating most frequently. The real-time profitability model (Figure 5.2) previously presented and discussed does this nicely.

This model implies that the frequently changing components of profitability are the production value of the products being produced; the cost of the

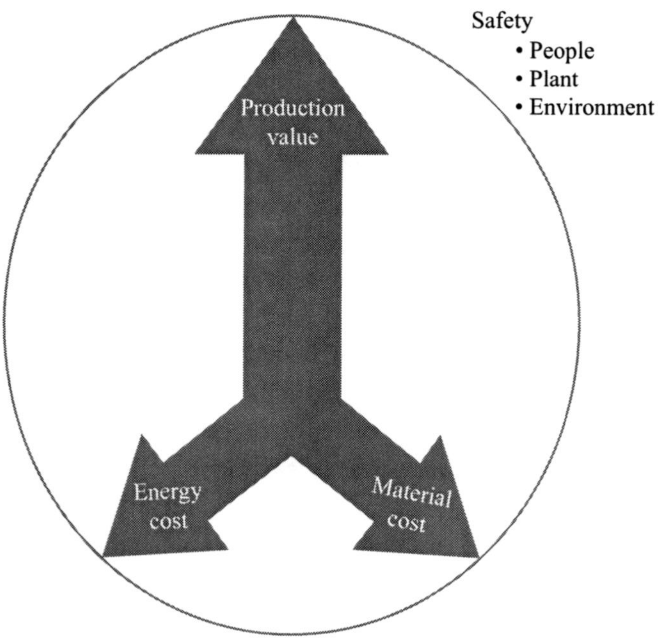

Figure 5.2. Constrained industrial real-time profitability model. Effective profitability control involves applying control theory to all components of the real-time profitability model: production value, energy cost, material cost, and safety.

energy being consumed; the cost of the materials being consumed; and the safety of the operations in terms of people, plant, and environment.

These are not the only components of profitability that need to be controlled, but they are the components that fluctuate in a submonthly time frame. Most of the other components of profitability are fairly stable over monthly periods and traditional approaches to managing them, using the monthly financial information produced by the Enterprise Resource Planning (ERP) systems still suffices.

For the sake of clarity, it is important to understand that the throughput, energy consumed, and materials consumed for industrial operations have fluctuated in real time for decades and still do. The primary objectives associated with controlling the efficiency of the operation are based on controlling throughput, energy consumed, and material consumed. Although the three variables that comprise real-time profitability—production value, energy cost, and material cost—appear to be the same as the variables associated with efficiency—throughput, energy consumed, and material consumed—they are not.

Production value, energy cost, and material cost are the economic value of production, energy, and material—not the quantity produced. This may seem like a fine point, but it is a critical point to the true understanding of the business problem faced by industry and the appropriate solution set.

The real-time profitability model clearly does not represent a comprehensive perspective of all of the components of the profitability of an industrial operation. Rather, it represents only those components that tend to fluctuate on a reasonably frequent basis—certainly less than monthly. Other aspects of the profitability of industrial operations, such as labor cost, do not tend to change very much over time. The month-to-month labor costs are typically fairly stable, unless an industrial company has a very difficult turnover problem. This is not to imply that those components of industrial profitability that do not change very frequently do not need to be controlled. They do. However, the traditional monthly reporting information from the ERP systems is typically sufficient in time and content for management to effectively control those fairly stable aspects of profitability.

The challenge has been that production value, energy cost, material cost, and safety, all tend to fluctuate many times every month. The monthly data provided by the ERP software have been proven to be insufficient for the effective control of these variables. Therefore, a new approach for the control of profitability is required, at least for these real-time components. It is interesting and helpful to keep in mind that the traditional approach for the control of profitability has been for managers to use the monthly data to determine current success and taking action based on the feedback the monthly data provide. In other words, the control system traditionally deployed for the profitability of industrial operations was a manual feedback control system, with business managers as the controllers.

There are two important characteristics to this traditional business management approach that were critical to its success. First, the information feedback was provided in a time frame that was consistent with effective decision support, even if that time frame was monthly. Second, since the component variables of profitability did not change frequently, a small number of managers had the time to make the profitability improving decisions for the entire operation. As some of the variables of industrial profitability transitioned to real-time variability both of these characteristics were driving profitability out of control.

The Profitability Control Loop

As was mentioned in the preceding brief review of feedback control systems, every feedback control loop has four basic components: a process, measurement of the controlled variable, a controller, and a controlling mechanism. In applying feedback control to industrial profitability all four components must be available and clearly defined (Figure 5.3).

The first component of the business control loop is the process. In good operating industrial businesses, the business processes are clearly delineated and well defined. It is very interesting that throughout the 1990s and early 2000s many industrial businesses invested considerable effort in business process design and business process redesign. It would appear as though the driving force associated with the need for better profitability control might have already been impacting many businesses. Business leaders with good intuition seemed to understand the need for doing a more complete and effective job

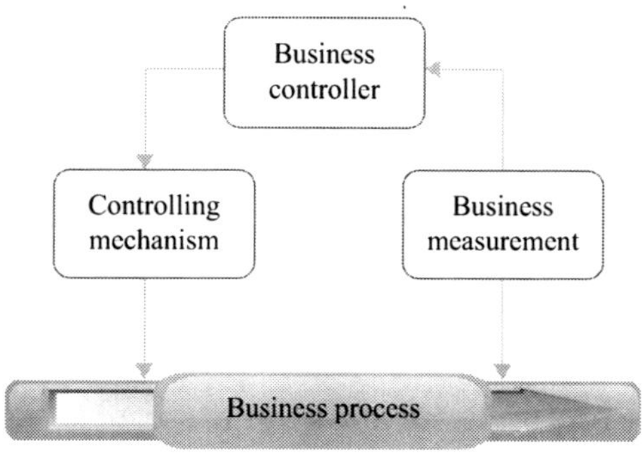

Figure 5.3. Profitability control loop. Similar to a process control loop, a profitability control loop has four components; a business process, a business measurement, a business controller, and a business controlling mechanism.

at defining and refining their business processes. Businesses that went through business process design or redesign activities find themselves in a much stronger position as they move to applying control theory to business variables than those who did not do this. In fact, the first step to preparing for real-time profitability control might be to do a structured business process mapping exercise. The workflow and structure processes associated with industrial operations and manufacturing are often quite clear and well defined by the process equipment used to produce the products. The business workflows may not be as visible, clear, or even as well defined. Clearly, working out the business process flows may take some effort, but the effort is certainly worth it in the long run.

One characteristic of many business processes that distinguishes them from industrial production processes is that of time constants. In many industrial processes, the speed at which the process can respond to an input is mediated by physics, chemistry, and biology. In the physical world, there are laws of nature that cannot be violated and determine how fast things might happen. Because these laws are predictable and repeatable we quickly learn how fast changes can occur in industrial processes. This speed is often referred to as the time constant of the processes and it tends to be a fixed value based on the laws of nature. Much of process control theory is tied to these naturally occurring time constants. Most business processes are not as well defined, repeatable, and mediated by natural laws of physics, chemistry, and biology. This means that the time constants of business processes are tied to human schedules and are, therefore quite variable. This characteristic becomes important when trying to apply control theory to these processes as we shall see a little further on.

The second component of the profitability control loop is the measurement. Traditionally, most business variables were measured on a monthly basis for the operation as a whole. Today, these monthly measurements are typically made in ERP software and are reported at the monthly close of businesses. As has been pointed out multiple times, this was sufficient as long as the key component variables of profitability change on greater than a monthly basis. As the components of industrial profitability in the real-time profitability model started to transition to more real-time fluctuation, the monthly measurement system was no longer sufficient. At a minimum these component variables of profitability had to be measured at a frequency equal to or faster than their rate of change. This provided industrial organizations with a huge dilemma, since there was no mechanism for measuring business variables on a reasonably frequent time scale and certainly no obvious real-time database to work from.

To exacerbate this dilemma, the industrial professionals traditionally responsible for managing the business information were the Information

Technology (IT) department, and the industrial professionals responsible for real-time computing and managing real-time information were the engineering department. In most industrial organizations, these two departments do not tend to interact with each other much. In many cases, since there is some functional overlap between these two groups, they do not tend to trust each other.[1] This has been a huge barrier to moving forward to real-time profitability control. Both of these groups have skills and knowledge needed to solve a part of this problem, but neither has all the necessary skills and knowledge. If these two organizations could join forces to apply their combined skill sets and knowledge bases to the problem of real-time profitability control, the solution might be at hand.

Engineers know that there is a real-time database in every well-controlled industrial operation that essentially conveys everything the manufacturing process is doing minute-by-minute and second-by-second. This real-time database is comprised of all of the sensors installed to measure the physical variables of the production process—flows, levels, temperatures, pressures, compositions, speeds, quality, and so forth. We do not often view these sensors as a database, but, in fact, that is exactly what they represent. IT professionals know how to access the key business variables associated with the profitability of the operations either internally from the business systems or externally from the suppliers and clients. Both data sources are absolutely necessary to construct the real-time profitability measurement system.

The responsibility for measuring the profitability of industrial operations falls to the accounting organizations. There are very specific requirements that define the way in which profitability and other business measures are to be determined. These requirements may change from location to location and often depend on the generally accepted accounting practices imposed by government agencies in each location. The accounting teams understand these nuances. Therefore, the accounting professionals must help drive the way in which the real-time profitability metrics are calculated. This presents another challenge to most industrial organizations. Accounting teams have traditionally performed their duties very independently from engineering teams. We (Peter and Walt) have been in a number of industrial plants in which the accounting talent and engineering talent do not even know each other. This must change for real-time profitability control to have any chance of becoming a reality. Engineers, accountants, and IT professionals must collaborate to develop and implement the correct real-time measures of industrial profitability. The resulting measurement system is referred to as the plant's real-time

[1] In the mid-2000s, I (Walt) ran a study of attitudes between plant engineering and IT departments. The most interesting comment came from a process engineer who said, "We get along just fine now—because I run both departments."

accounting system and is based on the concepts of activity-based accounting down to the unit or work cell level calculated in a real time frame.

Determining the correct real-time accounting measures is accomplished by performing a structured decomposition of the overall accounting measures of the operation. Today, monthly accounting is typically performed at a plant level for each plant in an industrial enterprise. These individual plant accounting measures are combined with those of other operations across the enterprise to produce the company's monthly accounting reports. The key to producing the correct real-time accounting measures starts with the monthly plant accounting measures and works down from there. This can be accomplished by clearly delineating the specific equations used by the accountants to determine the monthly financial reports. The accounting and engineering talent working together can then decompose the equations down to the plant area level by performing a contribution analysis for each equation to each area, that is, determining the contribution that an area makes each month to the overall financial measures of the plant. The result of this should be a contribution calculation for each financial metric for each plant area. These calculations can then be further decomposed down to the process units or work cells in the plant. A similar contribution analysis can be done for each equation at the plant area level. The net result of this decomposition will be a series of equations that define the real-time accounting models for each unit or work cell. The engineers who have a complete understanding of the real-time sensor database, the accountants who have an understanding of the business data required to make the calculations, and the IT professionals who know how to access the business information, must all collaborate to build algorithmic models for each of the financial equations at the unit or work cell level (Figure 5.4). Perhaps, the most challenging aspect of this real-time accounting model development is the translation of process sensor data, which is typically in a very specific engineering format, into the data elements required to populate the models. There is a bit of an art involved in this translation, but experience has shown that once the accounting and engineering talent build a level of trust and cooperate effectively, they can be quite successful at establishing the models.

Once the models are established on paper, they can typically be easily implemented in the controller environment of most available process control systems. Process controllers have been designed to execute algorithms for process control on real-time schedules. Most software for process controllers supports the easy development and execution of more general purpose algorithms. Since once the decomposition and modeling exercise are completed, the resulting models are typically equations or an algorithm representing the real-time accounting requirements at the process unit level, the process control software environment is ideal for the development and execution

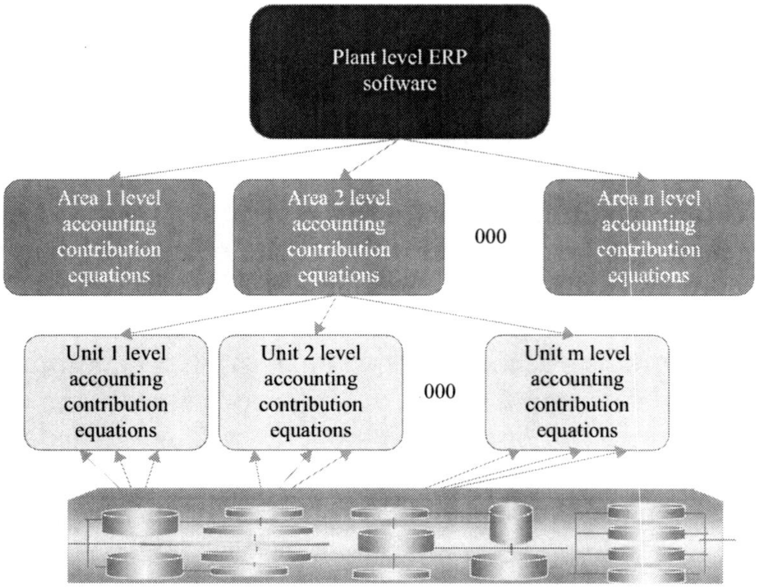

Figure 5.4. Accounting decomposition and modeling. An accounting decomposition is a top-down step-by-step process starting with the overall plant accounting algorithms and determining the contribution algorithms node-by-node to the process unit or work cell level.

of these models. In the case where a nonalgorithmic controller, such as a programmable logic controller (PLC), is deployed for the control function, the algorithms can typically be implemented in the human machine interface software that sits on top of the PLCs. In either case, the control system software provides the ideal environment for the development and execution of the real-time accounting models.

The idea of using the control software in automation systems to execute real-time accounting models is anti-cultural in most industrial operations. Traditionally, the IT systems were reserved for the business functions and the automation systems were reserved for the operations and engineering functions—and never the twain would meet. This traditional separation of functions proved very comfortable for industrial organizations that worked in effective silos. Each organizational silo had its well-aligned technology base for which they were responsible. The separation was clean. There was little overlap. It was comfortable. However, it stopped working well when critical business variables transitioned to real-time variability. To meet the new challenges, collaboration is required between both organizations and technologies. These cultural challenges may provide the largest hurdle to make the transition to the real-time industrial world. Overcoming this barrier will require real leadership clarity and strength. The benefits of making this transition are huge.

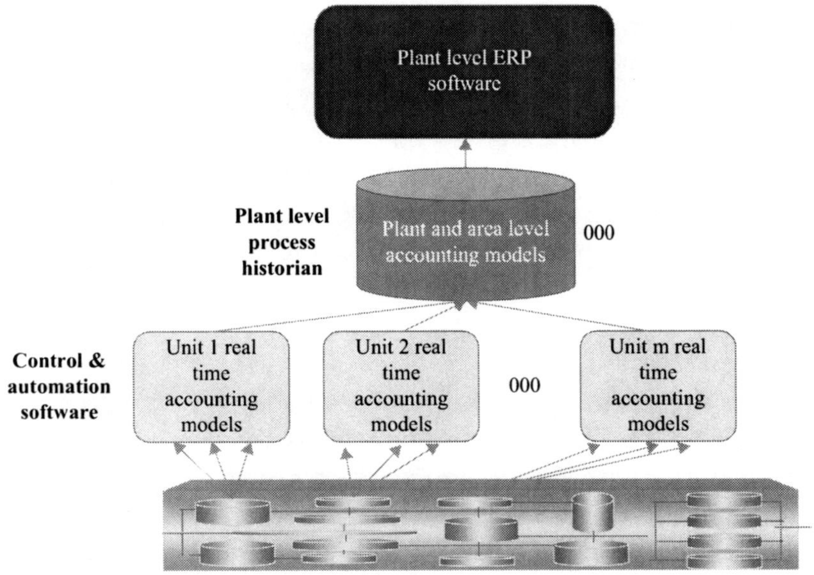

Figure 5.5. Real-time accounting system recomposition. When the real-time accounting system is implemented, it must flow from the bottom–up with the recomposition calculation at each node defined by the initial decomposition analysis.

Once the real-time accounting models are operating in the automation systems, the next step is to aggrandize them back up through the plant areas and to the plant as a whole (Figure 5.5). The basic system software necessary to implement this recomposition is readily available in most modern automation systems. This software is commonly referred to as process information management software or as a process historian. This software is designed to capture a history of basic variables coming from the process automation systems, but it is also designed to incorporate analytical capability that can be used to perform calculations and transformations on the incoming data. In the case of the accounting recomposition, the calculations and transformations required to take basic unit or work cell–level accounting data and compose it into the area level models should be obvious from the initial decomposition analysis. In some cases, it may be the simple sum of two lower-level metrics, but in most cases, it will be a more sophisticated transformation. The good news is that accounting, as a man-made discipline, tends to be very equation-oriented. Therefore, most of the transformations done by the process historian software will be based on fairly simple equations.

The area-level measures can be aggrandized into plant-level measures in much the same manner. This recomposed information, being managed in a process historian software environment, can also be stored over time in the historian. This allows for base unit accounting models, area accounting models, and plant-level accounting models to be historized on an hourly, shift, daily, weekly, monthly, or any other time basis may be of interest to plant management and accounting personnel. Once this information is stored in a process historian software environment it can be easily transferred to the plant and enterprise accounting system by linking the historian with the ERP software. In this manner, the real-time accounting environment provides a consistent bottom-to-top real-time financial and profitability system that has not previously been available to industrial companies. This is exactly what is necessary for the measurement component of the real-time profitability control loop.

The third component of the real-time profitability control loop is the controller. It is the controller's job to compare the current real-time business measurement values with the desired values and make a decision on how to move the current values toward their desired state. In process control systems, a simple general-purpose control approach has been successfully utilized for decades based on a proportional, integral, and derivative (PID) control algorithm. Correctly implementing and tuning a PID controller can be an engineering challenge, but the feedback control provided by these controllers can be very effective. It would be very nice if the same approach could be taken for the real-time profitability control, but, unfortunately it cannot be done. One characteristic of manufacturing and production processes that enables the use of the PID control algorithm is that the process loops have a fixed time constant defined by the natural laws of physics, chemistry, and biology. This time constant is actually used in the PID control algorithm. Unfortunately, business processes are driven by human schedules and not natural laws. As such, they do not have a fixed and repeatable time constant and the PID approach will not produce the desired results.

Determining the most effective control mechanism simply requires an analysis of early process control approaches as well as the process traditionally used to control industrial profitability. Early process control was provided by human operators. They viewed gauges to determine the current values of the variables within their domains of responsibility, and adjusted valves in the appropriate direction to control the process in a manner that would return the value back to the desired state. History has shown that operators were quite effective in performing this duty. Over time and with experience, operators learned how to control larger portions of the operation very effectively. The larger the scope of vision they had of the operation, the more they learned. The more they learned, the more effective they were. The point is that humans tend to be able to effectively control variables experiencing real-time fluctuation if they have a view of the measurement variables as they are changing, in real time, and contextualized down to the area of operation for which they are responsible.

In a similar way, prior to the transition toward real time of some of the component variables associated with profitability, business managers performed the basic profitability control functions for the overall plant. They could do this because the business variables changed so slowly that they had the time to make effective decisions. Also, the real-time operational variability was covered in the plant automation and control systems, which were designed to maximize the efficiency of the operation. The point is that humans seem to be able to do a very effective job at controlling the business variables of the profitability of industrial operations.

Combining these two experiences, it seems quite obvious that if the humans responsible for the performance of each domain of plant operation were provided the right business information on the right time frame, they could learn how to control the profitability of the operation.

Today's process operators are typically responsible for a fairly extensive logical domain of operation within industrial plants. They can manage this broad domain because the automation technology automatically responds to most process upsets and the operator's job has become one of oversight. Today's operator is also very different from operators of the past, having a much higher level of education, a much broader view of the operation, and often a much higher experience level. It seems logical that these process operators are in an ideal position to expand their responsibility level to move from process operators to performance managers by taking on the responsibility of profitability control as well as process control.

This expanded responsibility requires that the real-time performance measures necessary to make the appropriate, profitability-improving decisions, is provided to them in the right context and the right form and at the right time. Unfortunately, each process operator has a considerable level of responsibility and providing the right information in the right form requires very specific analysis. Since the operators are working in a real-time world, experience has demonstrated that they can be easily be overwhelmed if too much real-time data are provided and they are required to analyze all of the data in real time. Only the most important information should be provided in a simple format if the operators are going to be truly empowered to perform. Since operators are to be both responsible for the operational performance and the business performance of the operation, both prioritized operational and business performance measures are required to enable them to execute the necessary scope of work.

For many years, industrial companies have measured the operational performance of their operations by implementing key performance indicators (KPIs). KPIs are typically calculated on a daily, weekly, or monthly basis and are intended to show whether the operations is performing as desired. It is interesting to note that many KPIs are presented in a financial context, but typically without the sanction of the accounting departments, and therefore often have little credibility with business managers focused on profitability. But KPIs convey valid and important information about the operational performance of the plant and should be considered an important part of the performance measurement systems of industrial operations.

So fundamentally, there are two classes of performance measures that must be combined to provide the operators with the real-time feedback required to manage the operational and financial performance of their operating domains: KPIs and real-time accounting measures. As with the previous discussion associated with the real-time business measures, the first step in developing an effective performance measurement system for plant operators is to calculate all the measures, including the KPIs in real time for their domain of operation. This can be accomplished by performing a very similar decomposition analysis to that previously described for the real-time accounting measures. The KPIs can be decomposed from plant level to area level to unit or work cell level and modeled in the process automation systems. When both the operational and accounting measures have been modeled in this manner, all of the basic data for an effective real-time performance measurement system exist. However, there is just too much data to provide it all to the operators in real time. Therefore, a methodology that prioritizes the measures is required. Only the most important measures should be presented to the operators so that they are not overwhelmed.

The most effective way to prioritize the performance measures is to prioritize them to the current manufacturing or production strategy of the operation. Plant management established production strategies to respond to market and company requirements. The production strategy should be used to determine which performance measures should be considered to be most critical to the execution of the strategy for each operating domain. Dr. Thomas Vollmann of Boston University developed an effective analysis tool for the decomposition of performance measures to production strategy. This tool is referred to as a Vollmann Decomposition (Figure 5.6). This decomposition provides a simple approach to analyzing production strategy to performance measures. It is based on the Vollmann Triangle, which partitions strategy execution into three components. The first component that is at the pinnacle of the triangle is the statement of the strategy itself. The strategy should be stated in terms of specific actionable objectives. The second component of the model involves the development of an action plan for each of the actionable objectives stated in the strategy. Each action step in the action plan should be measurable. The third component of the model consists of the measures of success of each action plant. These measures are referred to as strategic performance measures.

This analysis should first be undertaken for the entire industrial enterprise. If an industrial organization is partitioned into divisions, each divisional

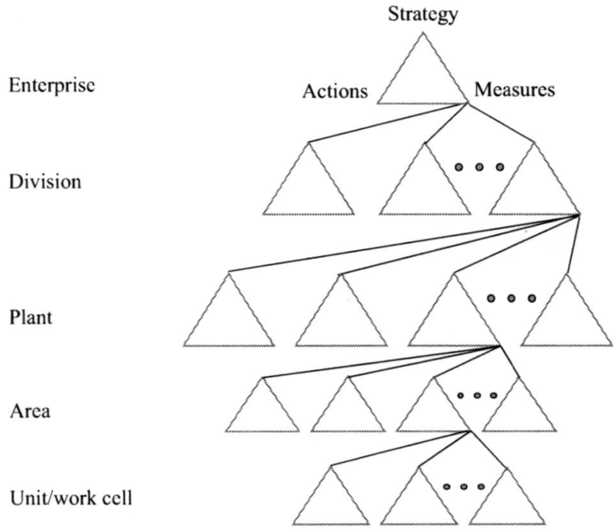

Figure 5.6. Vollmann Decomposition. Dr. Thomas Vollmann developed a structured top-down approach to the analysis of manufacturing or production strategy that enables the action plans and strategic measures supporting a strategy to be identified and prioritized right down to the process unit level.

management team should evaluate the enterprise-level performance measures and develop a Divisional strategy, action plan, and set of performance measures based on the Divisional contribution to the Enterprise-level strategy. Once this is complete, each plant management team in the enterprise should evaluate their contribution to the performance measures for the enterprise and determine the plant's strategy, action plan that performance measures. Then a similar analysis should be executed at the plant area and process unit or work cell levels in the plant. The result will be a set of performance measures prioritized to strategy for each process unit. Done correctly, these performance measures should be subsets of the KPIs, and real-time accounting measures. The net result is a set of performance measures, modeled in real time and prioritized to the manufacturing strategy. This prioritized set of real-time performance measures is commonly referred to as the Dynamic Performance Measures (DPM) of the business.

Once these DPMs are modeled and available, they need to be aligned to the domains of operation for each operator. This is typically fairly easy to accomplish since most operating domains are logically assigned according to process workflow, which is aligned to the specific equipment in the plant. This means that each operating domain can typically be defined as a process unit, set of units, area, or set of areas. Therefore, the contextualized measures for each operator should align nicely to the nodes of the Vollmann Decomposition. Since the operators are performing their tasks in real time, and since performance measures often compete with each other, it is very important to limit the number of DPMs provided to each operator. Experience and ergonomic studies have shown that limiting the performance measures to four or less provides the best opportunity for the operators to perform well.

The feedback mechanism for the operators should be a simple dashboard or scorecard display. Since the visualization capability of computerized automation systems is so advanced, there is a tendency to want to build visually striking dashboard and scorecard displays, but experience has shown that the simpler the better—as long as all the necessary performance information is available. Operators perform most effectively if both the current value and trend history of each performance measure is displayed. This allows the operators to evaluate current performance and to see if past actions are having the desired impact. Contextualizing the dashboards and scorecards to the operators' domain of responsibility is as simple as mapping the measures to the operators' assigned domain (Figure 5.7). Since the domain of responsibility of each operator may be larger than a single unit, the measures the operators typically have displayed on their dashboards are of their entire domain. As one

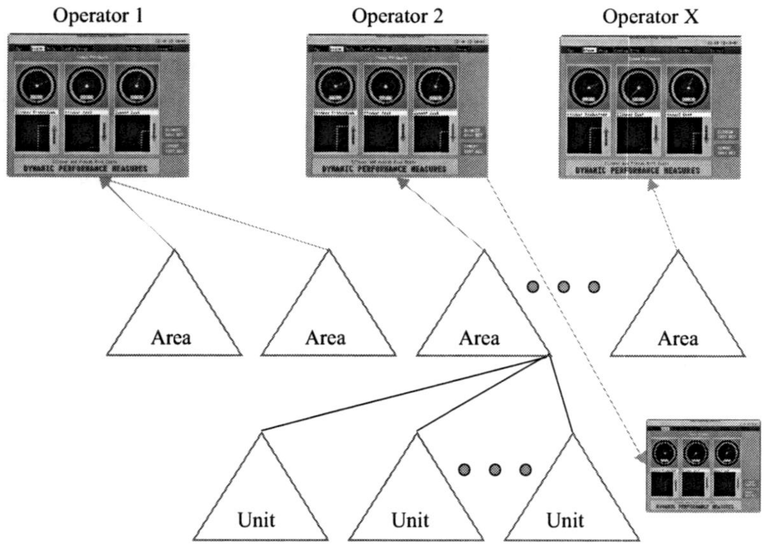

Figure 5.7. Operator alignment and drill down. Executing a Vollmann Decomposition analysis right down to the process level and contextualizing the strategic measures to an operator's domain of responsibility ensures overall strategic alignment. The structure also provides for detailed drill-down displays within each operator domain.

or more of the DPMs starts changing value, either in a good or bad direction, it may not be apparent to the operator what the underlying reason is for the change. Providing a drill-down mechanism so that the operator can view the DPMs of the process units or work cells comprising their domain can be very useful. The operator can drill down to the lower level, see what is going on there that may be causing the change, and learn how to respond most effectively. This visual empowerment of the operations staff is really what transforms them from process operators to performance managers.

Since managing the profitability of the operation is the responsibility of not just the operators, but of every person in the operation that could impact profitability, once the operators' dashboards are implemented, it is very helpful to implement dashboards for the others (Figure 5.8). The information provided on each dashboard or scorecard needs to be contextualized to the specific responsibilities of each function. Doing this will encourage and empower each profit-impacting person in the operation to perform his or her responsibilities more effectively and to align with every other person. An environment that encourages cross-silo-collaboration will result.

The final component of the profitability control loop is the controlling mechanism. In the case of process control the controlling mechanism

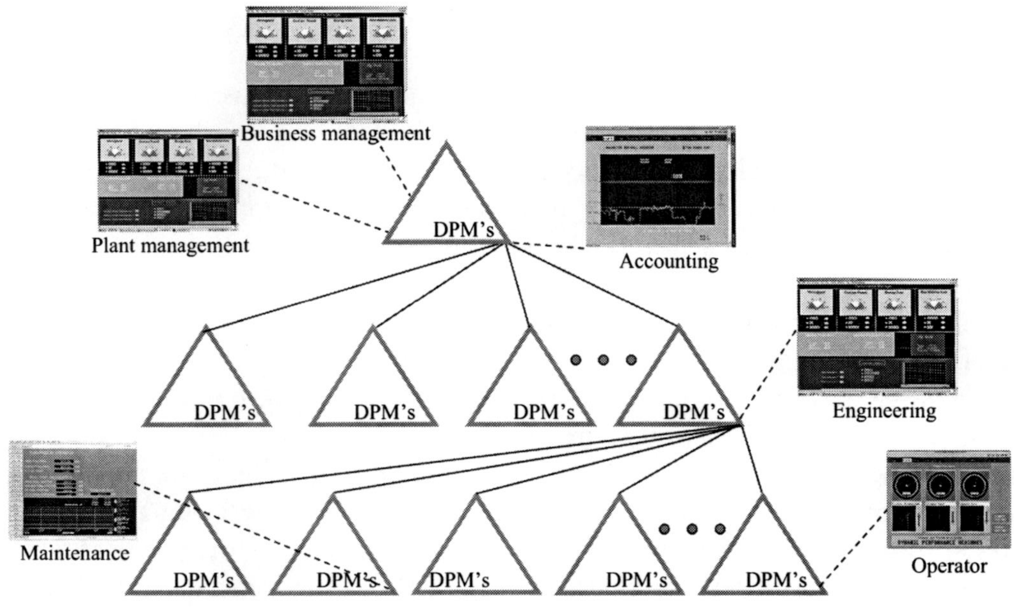

Figure 5.8. Organizational empowerment through DPM visualization. Providing each person in the operation with a real-time DPM dashboard contextualizes the jobs they perform and creates an aligned empowerment approach in which the actions of each person should complement those of all others.

is typically a valve, pump, or some similar device that directly modifies the process to realize the desired response. In the case of real-time profitability control, the controlling mechanism is not quite as obvious. To understand the basis of the appropriate control mechanism, it is important to understand that prior to the empowerment of the process operators by providing them with real-time decision support information on real-time dashboards, they were already making hundreds of decisions each day that impacted both the efficiency and profitability of the operations. Each time an operator changes a set point or a working recipe variable, for example, the change cannot be neutral. It either creates business value or destroys it. Typically, each decision does not have a huge amount of impact, but since the operators are making many decisions each day, altogether, the overall impact can be very significant. The point is that prior to the availability of the real-time business decision support dashboards the operators did not know whether each change they made was good or bad for the profitability of the operation. Good operators want to do what is good for the operations, but without knowing what is good, they make their best guess.

With all of this in mind, the basic controlling mechanism for real-time profitability control is merely the operators doing what they have always done, but doing it with additional real-time feedback on how their actions impact profitability. In essence, the operators really do not do anything different from what they have always done, but they do it differently. They make their decisions with real-time profitability feedback. This makes a huge difference to the profitability of the operation.

This is a very important concept for engineers and industrial managers to understand. As has been previously discussed, many industrial engineers and managers have developed a distrust of their operators due to the traditional characteristics of the labor force in industrial organizations. Expanding the functional scope of what frontline operators do can be very uncomfortable to engineers and managers. It is important to realize that this approach does not really involve increasing the functional scope of what operators do. Rather, it increases the intelligence they have in performing their traditional functions. Sometimes, understanding this subtle point makes the cultural transition from operators to performance managers easier to accomplish.

The operators are not the only personnel who participate in the control mechanism of the real-time profitability control loop. Every function in the organization whose actions and activities have an impact on the profitability of the organization also becomes part of the control mechanism. This is done much in the same way as for the operators, by providing contextualized real-time feedback through appropriate dashboard and scorecard displays. Once again, the basic job functions of each person remain the same, but they perform their functions making better decisions for the profitability of the operation.[2] As each function makes better decisions, those decisions contribute to an overall improvement in profitability. The profitability of the operation is brought under manual control, much in the same manner as the process was brought under manual control through process operators during the early phases of the Industrial Revolution.

As real-time profitability-based decision support moves down through the industrial organization, the business managers of the operation must feel comfortable delegating performance management to the lower levels. This means that the business managers are not the sole employees responsible for controlling the profitability of the operation—everyone is. Therefore, the job of the business manager must change to more of a management oversight role rather than a direct profitability control role. Initially, this can be very uncomfortable for the business managers because it may feel as though they are giving up responsibility—and they truly are to some degree. However, as the business variables continue toward real-time fluctuation, it will not take much time for these business managers to realize that they cannot control profitability any more without help. Things are just moving too fast for this to be the case. Delegating this responsibility may be a bit uncomfortable, but it is absolutely necessary for binging profitability into control.

A good sports analogy might be appropriate. I recently watched a group of fairly young children playing baseball in an instructional league. I noticed that as their coaches were trying to teach them how to hit a pitched ball, they would often call out to the batters when to swing the bat. This seemed to have been a very useful instructional approach. Much of the decision making on when to swing was with the coach. If this continued as the players developed it could be a great hindrance to success. For example, if the coach of a professional team still made the decisions as to when the batter would swing and at which pitches they would swing, the batter's effectiveness would go down significantly. To optimize success, the coach has to give up responsibility to the batter who is in a better position to make real-time decisions than the coach. This is very similar to what needs to be done between the business managers and lower-level functions in industrial organizations. Transitioning all profit-impacting personnel into performance managers is the key to developing the most effective controlling mechanism for real-time profitability and completing the real-time profitability control loop.

With time, automatic control response may become effective in profitability control, but for the most part, manual control will be most effective in

[2] For years, it was simply assumed that each control loop and each unit in a process plant needed to be optimized so that it performed to its maximum tunable capability. When plants are operated in the manner we are advocating, it may actually require that a loop or a unit may need to be detuned to keep it in line with the rest of the operation, as suggested by Eli Goldratt's description of a "Herbie."

the initial stages of transition to real-time profitability control. It is difficult to envision how a simple repeatable control algorithm, such as the PID algorithm of process control, may be developed for profitability control. One technology that seems to offer promise is automatic workflow software. A workflow can be triggered from a change in profitability or by some other relevant event. The preconfigured workflow can then analyze the event to determine an appropriate way to respond and then can actually automatically respond. As more experience is gained in applying real-time profitability control, it is expected that a stronger perspective on how these loops may be automated will evolve.

One of the biggest stumbling blocks to enterprise integration has always been time base. As we have explained, real-time control operates in fractions of seconds, while transactional control operates in days, weeks, or months. Workflow is time base nondependent. Experiments are being done by the Smart Manufacturing Leadership Coalition among others to determine the feasibility of expressing all control loops from the business down to the actual sensor PID loop as workflow.

A CASCADE CONTROL SYSTEM

It is important to put real-time profitability control for industrial operations into a little perspective from the point of view of control theory and, in particular, from the point of view of industrial process control. As the concept of real-time profitability control started to evolve it was often viewed as a replacement for traditional process control. This is simply not the case. Real-time profitability control can work reasonably only if the process control system works effectively and efficiently. In fact, it may be helpful to view the relationship between profitability control and process control as a cascade control strategy, with profitability control as the primary control loop and process control as the secondary loop (Figure 5.9). This is a bit of an oversimplification because there are clearly multiple control loops involved with both process control and profitability control, and the relationship between profitability control and process control will most likely not be a one-to-one relationship. However, in concept, it may be helpful to consider the overall structure in this manner.

Viewing real-time profitability control as the primary loop of a cascade control strategy has the benefit of exposing some of the requirements for success. The objective of the process control systems is to maximize the efficiency of the industrial operations. Well designed, implemented, and tuned process control systems are very effective at controlling plant efficiency. It is important to understand that the process control system must be performing effectively prior to implementing a profitability control strategy. This is very much in line with the requirement that the secondary control loop in a cascade control

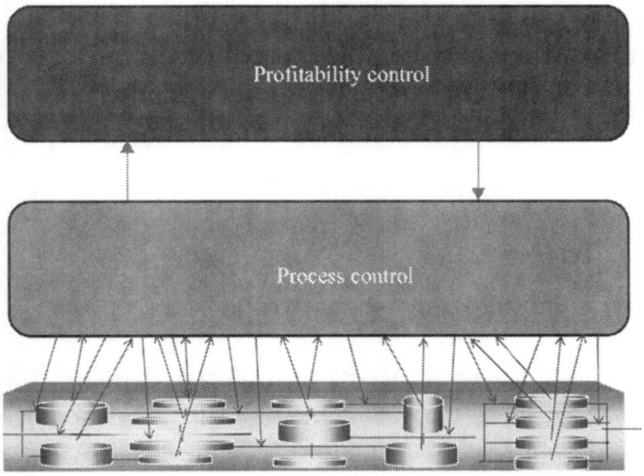

Figure 5.9. Profitability control cascaded to efficiency control. Perceiving the relationship between profitability control and process control as a cascade control strategy is helpful when developing effective holistic control strategies.

strategy must be working effectively prior to connecting it to the primary loop. If this is not the case, the primary loop will have little or no effect. Therefore, a well-operating process control system is absolutely a prerequisite to real-time profitability control.

A second characteristic of the real-time profitability cascade control strategy is that the time constant of the secondary loop—the process control system—must be considerably shorter than that of the primary loop—the profitability control system. In most process plants, this is the case, especially since governmental agencies regulated the frequency of a number of key profitability control variables, such as the price of electricity. If the process control system cannot respond fast enough to required changes in the profitability control system, the overall system will continue to be out of control to some degree. Take the example of an extended pipeline carrying crude oil to refineries. The price of the crude oil may change many times over the duration of time it takes to ship the oil through the pipeline. There is very little the company that owns the oil can do to control the profitability of the crude as it makes its way through the pipeline. In this instance, therefore, although the pipeline process may be in control, it reacts too slowly for the profitability loop to be in control. Each circumstance of this nature must be evaluated on its own characteristics. In the future, this may present industrial companies with significant challenges on how to restructure their basic processes to bring them under better profitability control. In the example pipeline process, the cost in terms of unrealized profits may become so onerous at some point in time as to cause the company to consider building a refinery close to the source of the crude. By doing this, they may be able to reduce the process time constant to the point that the profitability can be brought back into control. Naturally, the benefits from such a decision would need to be considerable to justify such a capital expense.

THE NEED FOR ENTERPRISE CONTROL SYSTEMS

The concept of real-time profitability control is fundamentally simple, but the installed IT and automation infrastructure in most industrial enterprises provide a barrier for the effective implementation of a plant-wide or enterprise-wide real-time profitability control solution. For a variety of reasons the automation and information systems installed in most industrial plants and across most industrial enterprises have been installed as discrete islands of automation and information. The result is that most plants have a number of automation and information systems in place from a number of different suppliers and they do not tend to work together. Many of these systems were initially designed around functionally specific domains, such as Distributed Control Systems (DCS) for process control; Safety Instrumented Systems (SIS) for safety control; PLCs for discrete logic control; Laboratory Information Management Systems (LIMS) for laboratory information management; Manufacturing Execution Systems (MES) for planning, scheduling, and other plant operations management functions; and ERP systems for financial reporting and management. Unfortunately, a prerequisite to a comprehensive real-time profitability control system across a single plant is that the different automation and information systems interoperate in a manner that enables a holistic approach to the business of the plant. This is certainly also true for interplant operations across entire industrial enterprises.

The cost to interconnect the different systems together in a point-to-point manner, which is about the only possibility for bridging the islands, is typically prohibitive. Although most vendors of intelligent systems provide software, most of this software is designed to connect to a single model of system or technology. The result is a nonsystematic and high-cost entry point for enabling a real-time profitability control approach. Even if an industrial company decides to try to build a plant-wide or enterprise-wide system using this point-to-point approach, the resulting system would not work as a single system and would require such a level of system-specific knowledge and skills that the result would be a very nonhomogeneous, and very difficult to use system. Not to mention that when each system is changed, due to reprogramming or upgrade, the application programming interface (API) between the system and the rest of the enterprise being "brittle," will break, and require new programming necessitating the control loop to be run in manual mode until the API once again is operational.

Fortunately, a new class or real-time system has begun to emerge over the past few years offering great promise in providing the basis for real-time profitability control across plants and entire industrial enterprises. This new class of system is referred to as Enterprise Control Systems (ECS). ECS are designed to unify the traditionally disparate real-time information and control domains into a single, systematic real-time computing environment that can effectively interoperate with the ERP system software to incorporate the traditional business management functions (Figure 5.10).

There have been three basic approaches to the design of ECSs. First is to design the ECS from inception to incorporate all of the functions of the traditional real-time automations and information systems, such as DCSs, SISs, PLCs, and MESs into a single real-time environment. In this case, the industrial companies can replace aging systems that need to be replaced with ECS replacement components as they hit their obsolescence point. Over time, as all of the older systems are replaced, the industrial company will have the unified ECS environment required to enable a real-time profitability control approach.

The second approach is for the supplier to design an ECS system that provides a supervisory software layer interfacing to the installed real-time systems

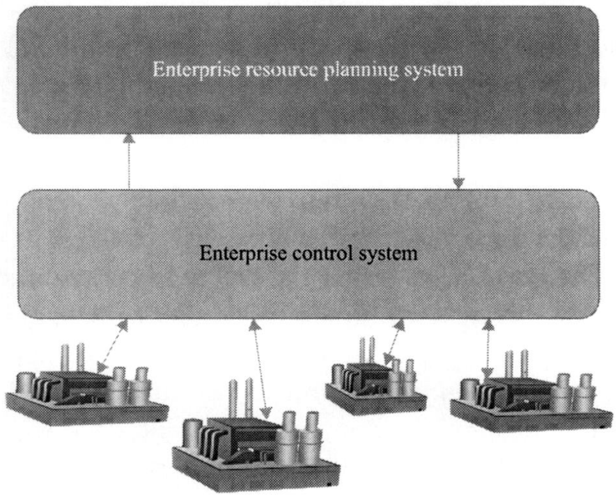

Figure 5.10. Enterprise control systems. Over the past two decades, the business-level functionality has merged into fully functional Enterprise Resource Planning (ERP) systems through service-oriented architectures. A similar convergence is occurring at the operational level to form a new class of automation systems—Enterprise Control Systems (ECS).

and technologies to provide a unified real-time system environment incorporating all of the installed systems. This requires that the ECS designer have the capability to design specific interfaces to all of the systems installed in industrial plants, whether open or proprietary. The advantages of this ECS approach is that an industrial company can move toward a unified ECS without the huge capital cost involved in replacing all existing systems and can typically get to a unified ECS environment in a much shorter time frame. The disadvantage over the first approach is that the resulting system, being comprised of multiple different systems from multiple different suppliers, will not be quite as unified as is a single supplier ECS.

The third approach is a hybrid approach that combines the characteristics of the previous two approaches. In this case, the ECS infrastructure can be installed over the existing automation and information assets to provide an initial unified real-time environment at a reasonable cost and in a reasonably short time frame, and then expand to become a single-vendor system as each component reaches end-of-life. This requires both open connectivity and a functionally complete ECS that incorporates the necessary functionality from the previously distinct classes of a real-time automation and information system.

Regardless of which approach an industrial company might take to move toward a unified ECS environment across each plant and the enterprise as a whole, as this is accomplished, it enables the implementation of a real-time profitability control solution. The more of the enterprise that is encompassed into the ECS environment, the broader the real-time profitability solution can be. As more and more of the company transitions to an ECS approach and real-time profitability control, the resulting profitability gains, which will be measureable due to the measurement component of the real-time profitability approach, will prove to more than justify the expenditure.

As industrial environments transition to real-time variability of some of the key business variables impacting their profitability, business as usual is just not a viable option. The good news is that the ability to control the profitability of industrial operations through real-time profitability control exists and can help meet the daunting challenges faced by industrial companies. Now with the emergence of ECS technologies the unified real-time system base required to implement real-time profitability control exists. The challenges may be great, but the solution is at hand. The real-time industrial business environment can be brought under control by the effective application of feedback control theory.

6

The Barriers to Real-Time Enterprise Control

The basic concepts associated with controlling the real-time enterprise are quite straightforward. Since industrial enterprises are typically complex, developing a holistic real-time enterprise control system can be equally complex. As with any highly complex strategy, developing and utilizing structured analyses and work plans are essential to ultimate success. Unfortunately, as with almost any organizational or system transitional process, there are barriers to successful implementation. Some of the barriers are organization specific, but there are a few that are common to every industrial organization and that can be expected to be encountered. Sometimes, merely understanding that a barrier may exist and is likely to be encountered is a critical first step in overcoming it. It is our intent in this chapter to expose some of those common barriers to success and suggest potential strategies for overcoming them.

Experience has demonstrated that five cultural and organizational barriers to successfully implement a real-time enterprise control system are commonly encountered and should be expected. They are:

- A Technology versus Solution Focus
- The Capital Project Process
- Traditional Accounting Practices
- Common Engineering Perspectives
- Organizational Silos
- Business and Production Process Limitation

Each of these represents commonly accepted norms in industrial organization that have underpinned traditional business success, but can derail the transitions to a real-time enterprise control system. The history and background that resulted in each of these barriers, the problems the barriers create, and a strategy to overcome each barrier will be worth looking at in detail. Effectively addressing these barriers is essential if the full benefits associated with controlling the real-time enterprise are to be realized.

A TECHNOLOGY VERSUS SOLUTION FOCUS

Technology in some form or another has been used to monitor and control industrial processes almost since the inception of the Industrial Revolution. In early industrial operations, the technology might have been as simple as an electromechanical relay to get a machine started, or a sensor and associated gauge to convey the temperature in an oven. As industrial factories and plants increased in sophistication and complexity the technology used to control the manufacturing processes increased as well. In fact, in many instances, the increase in sophistication of the instrumentation and control technology actually enabled the increase in sophistication and complexity of the production processes. The scientists and engineers who developed the instrument and control technology had in-depth understanding of the manufacturing and production processes. They understood what the industrial companies were trying to accomplish, the challenges and constraints they were facing, and the impact those challenges and constraints were imposing. Instrument and control technologies were specifically developed as solutions to meet those challenges and overcome those constraints. In this environment, the instrument and control technologists were truly solution-focused.

Both of us recall that when we first entered the field of instrumentation and control systems in the 1970s, the instrument and control companies worked hand-in-hand with the process engineers in the industrial production companies to understand the pressing process issues and problems and develop technological solutions. Instrument sales people typically were professional engineers and carried overalls and extensive tool kits in their cars so that they could work with their customers to build control solutions to solve their problems. If the challenge proved to be too much for them, they would call back to the main office and engage instrument and control experts. Together, this team of professionals would design and develop often very sophisticated solutions.

At that point in time the instrument and control industry as a whole was much more focused on solving manufacturing and production process problems than on the technology underpinning the solutions. The technologies that made the solutions possible were seen to be enablers of the solutions. It was the knowledge and capabilities of the instrument and control engineers that was valued throughout industry. The field of instrumentation and control was truly a solution-focused field.

A similar analysis can be made of business management. Prior to the 1960s, the disciplines associated with managing business functions were led by expert problem solvers in each discipline who were responsible for using the tools of their trade to effectively meet the needs of the business. Teams of accountants would perform the accounting functions, financial analysts the financial analysis functions, payroll specialists the payroll functions, and so on. Business was organized around functional expertise with the tools to perform their functions. Each area was solution focused.

In both the business management and operational sides of industrial businesses the solution focus seems to have changed almost overnight. The catalyst for the change was the introduction of the digital computer as the primary technological tool for performing both business management and process control functions. The computer was introduced as a tool for business management functions a bit earlier than for process control, but the impact on both was very similar. The computer was so complicated in its own right that industrial companies had to hire experts in computers to get them to work as desired. On the business management side, Management Information Systems (MIS) departments were built with computer experts who did not typically understand the business management functions but did understand aspects of how computers worked. System analyst positions were created to try to translate the business requirements into computer terms so that the computer programmers might be able to develop the programs to perform the functions desired. The MIS team was technology focused not solution focused. This caused great difficulty for business management functions for decades as the business functions continually complained that the programs did not work as they were supposed to and the MIS team continually responded that they worked as specified. At times, the tool itself became a barrier to successful business management functions.

A similar shift occurred in the process operation side of industrial businesses. As computers became less expensive with the introduction of the minicomputers in the late 1960s and 1970s, it became practical to start to use them for process control. Unfortunately, most instrument and control engineers did not understand the computer and had difficulty transferring their solutions knowledge into computer terms. Instrument and control companies started hiring computer knowledgeable talent to help better utilize the technology. Unfortunately, most of the computer professionals did not have a good understanding of the process management and control. Some instrument and control companies actually organized around the technology with an analog division and a digital division. The analog division typically focused on instrument and control solutions utilizing analog technologies, while the digital division typically focused on the computer technology.

Since advances in digital computer technology occurred very frequently, most industrial and instrument and control companies had to invest considerable talent and effort into keeping up with the advances and trying to determine how each advance might impact industrial plant and process operations. The primary focus of industry often shifted to the advances in technology and how those advances might be pushed into industry rather than on solutions to the production challenges faced by industry. The instrument and control engineers who understood the production challenges started to take a back seat to the automation engineer who understood the technology. Industrial automation became highly technology focused and much less solution focused.

There are a large number of examples of technology-focused initiatives and programs that have impacted industrial companies over the past three decades. One of the most obvious and interesting examples was Computer Integrated Manufacturing (CIM), which became a dominant focus of industrial automation in the 1970s and 1980s. Another example of a technology-focused movement took place during the late 1980s when expert system technology was driving some excitement in the industrial marketplace. Expert systems appeared to be exotic and were seen for a while to be the solution to any problem that was not previously solvable. Every industrial automation vendor introduced products that were based on expert system technology. The industrial market bought these products like they were going out of style. It did not seem to matter whether a problem was algorithmic or heuristic in nature, if a supplier had an expert system-based application directed at the problem the market would buy it. This is another classic example of the technology being the focus rather than the industrial problem. Expert systems can provide excellent bases for a variety of solutions, but when the technology overshadows the solutions it can lead to less-than-satisfactory results.

Another interesting example that was taking place about the same time as the expert system craze was lights-out manufacturing. The idea behind this initiative was that technology had advanced to the stage at which entire manufacturing and production operations could be completely automated and people would no longer be required. If there were no people there would be no need for lights, hence the term lights-out manufacturing. This was a fairly short-lived movement as industrial companies quickly realized that they could not automate for every contingency and that people were necessary. I recall one operation manager joking that he would go to lights-out, but his operators may not be able to find their way to the control room.

The move to Manufacturing Execution Systems (MES) followed a similar technology-focused path. The first time I heard the phrase was at an Advance Manufacturing Research (AMR) conference in Boston in the 1980s. The president of AMR, Tony Frisca, was doing a presentation on the separation of systems into business information systems and automation systems, but he pointed out that considerable software applications were being developed by a variety of suppliers in the space between the business and automation systems (Figure 6.1). Since the application development in this middle area was not being done in any structured manner, he declared the middle space to be a mess. He went on to say that the industry standard for acronyms was three letters; so he decided to label this middle space MES for the mess that it was. After establishing the label he came up with the phrase MES. Ever since that day, both industrial automation vendors and industrial companies have been trying to develop and implement the best MES they can. Even in the 1980s, Tony was trying to point out the technology focus of industry—with a bit of humor.

There are many other examples of technology-focused approaches and movements that have impacted industrial automation over the decades. It is important to realize that none of these was bad and each had a real basis of value. It is just that they were focused on the technology, not the problem the technology exists to solve. Today we are still very much in a technology-focused market. Movements such as Wireless and Cloud Computing follow the same technology focus as their predecessors. Since industrial automation solutions are often enabled by computer technology, industry must work to understand the value of new technological approaches as they become available.

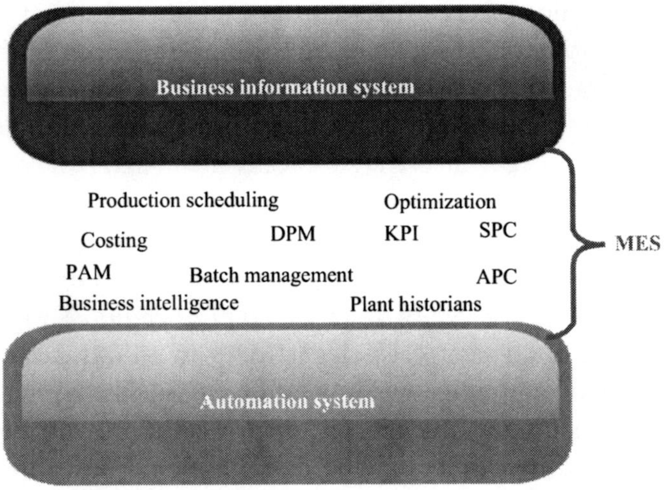

Figure 6.1. Manufacturing Execution Systems (MES). The labeling of the space between business and automation systems as MES because it appeared to be a mess was a humorous but accurate portrayal of an example of the technology focus of industrial automation/information systems.

However, the ease-of-use and broad capability of computer-based technology have reached the point at which the predominant focus of industrial automation companies and industrial production companies can transition back to an industrial solutions focus.

Controlling the real-time enterprise is a prime example of a solution-focused approach. One major challenge industrial companies have to face is that the business of industry is moving toward real-time variability and the systems in place are not sufficient in dealing with this. Continual technology-focused initiatives will not get industry to where it needs to be to meet this challenge. It is time to start focusing on the real-time operational and business control problems industry is facing and start dealing with them in the tradition of the great instrument and control engineers of the past. Industry is facing a real-time control crisis, but has all the tools and talent to meet this crisis head on by taking a solutions-focused approach.

THE CAPITAL PROJECT PROCESS

A second barrier to the successful approach to controlling the real-time enterprise is the traditional capital project processes utilized in industrial operations. Most industrial companies partition their budgeting into expense and capital budgets and have strict rules on the development and deployment of each budget. These rules have been critical to the successful financial management of industrial companies for centuries and are highly ingrained into their cultures. Seldom do industrial companies question the wisdom of these processes and associated rules, so it is not always obvious when some of the traditional approaches are hindering progress. Both Information Technology (IT) and industrial automation expenditure are typically governed by the capital project process of industrial organizations.

Most industrial companies have highly structured capital budget and project processes that prioritize and govern the use of capital budgets. Figure 6.2 provides an overview of the major steps of a typical capital budget and project process. Typically, once a year, each department in an industrial organization puts forward a list of capital expenditure they would like to make. As part of each request they estimate the expected return on the investment (ROI) for each of the proposed investments. The cumulative requests are presented to management that evaluates and prioritizes each request with respect to company plans and directions and the expected financial returns. Management then determines which of the prioritized capital requests they can afford to fund within the limitations of the available capital. Those that make the cut are funded and those that do not are turned down but might remain in the queue for later consideration.

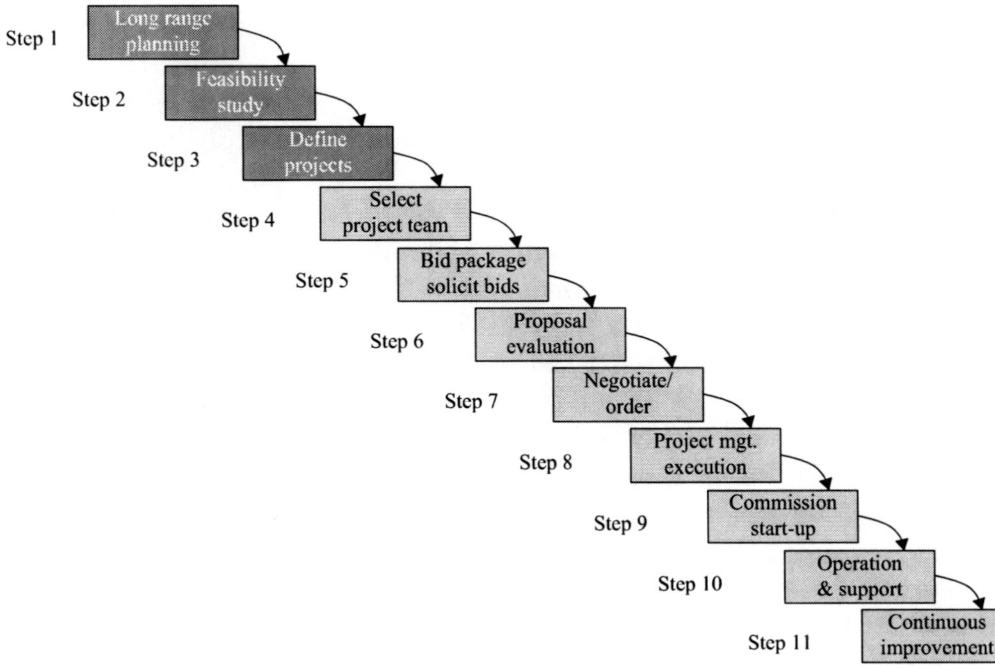

Figure 6.2. Sample capital project process. The process for allocating capital to projects and executing capital projects was necessary for managing capital requests, but it tended to suppress the value potential from many automation projects.

When a capital request is funded, an internal project team is typically assigned to oversee the execution of the capital project. The project team is responsible for developing the project schedules, defining the project requirements, selecting the product vendors, selecting outside services that may be required, managing the execution, and delivering the project. The measures of success of the project teams are almost exclusively the *on time* and *on budget* delivery of the project. Once the project is delivered, the members of the project team typically either go back to their regular job or move on to the next project. Their mission with respect to the completed project is essentially over. If the project was completed on time and on budget, the project team is considered successful.

This type of capital budget and project process has been successfully used in industrial companies almost since their inception. However, there are certain aspects of this process that can hinder success when moving to driving the *increased value* that comes from controlling real-time enterprises. The first aspect is that project teams are seldom measured on the value the project generates for the business. They are almost exclusively measured on on-time, on-budget delivery of the project—whether or not the result truly adds any value. This fundamentally means that decisions made during the project cycle are primarily determined by their impact on project delivery of the project and not the incremental business value in which they may result. This can be very frustrating for automation and IT providers who have value-generating technologies and solutions. The value-generating aspects of their solutions are almost never considered by the project team, especially if these solutions add any time or cost to the project. Since the project team is not measured on value generated they tend to select technologies that are easy and fast to implement over technologies that add significant incremental value. For example, there are two possible technical solutions to a project definition and the first potential solution has a feature that will increase the annual profitability of the operation by $1 million over the second; the project team may be likely to select the second solution because it will help them to reduce project cost. This means that the ongoing business operation would be stuck with a solution that does not meet their business objectives as effectively. The measures of success of the project teams are typically not aligned with the measures of success of the business. The net result is that technology vendors have to focus on low price and ease-of-use over value in order to win the business. Since transitioning to a system to control the real-time enterprise in order to increase the profitability of the operation often involves increased scope of functionality over traditional approaches, if that increased scope is not carefully included in the project definition it will not make it into the delivered solution. This is a risk when moving to any new, value-generating solution or technology.

In the specific case of automation and information investments, which are by nature very complex, the process commonly followed to develop the functional definition of the project only adds to the difficulty. Many automation and information capital project requests involve updating existing systems. Project teams do not tend to develop a detailed functional requirement specification until after the capital projects have been approved. To effectively deal with the mechanics of the capital budgeting process, the initial specifications developed when applying for capital investments are often based on the specifics of the technology that is currently installed and is to be replaced. This means that any new value-generating function or solution that may have been developed since the technology was last installed will not be included in the initial specification, and the budget request will be made without including the new feature or solution. When the budget is approved, it becomes very difficult for the project teams to include the new value-generating features within the initial requested cost and scope.

To further exacerbate this situation, although there is a need to predict the expected ROI for each capital investment, seldom is the value from the investment measured once the project is complete. The cost accounting systems installed in industrial operations are typically based on monthly financial reporting across the entire industrial site. Most capital projects impact a small subset of the overall operation. Even if they are adding value, the increase in value will almost never be discernible by the accounting system. This means that management never knows the true value of most automation and IT projects to their ongoing operations. Unfortunately, after a number of capital projects with no discernible value improvement, management often develops a jaded attitude toward this expenditure. Many believe that if the value is not measureable and discernible, it does not exist and future expenditure goes to other areas for capital investment. Hence, the Enterprise Control System could well be measured against the cost of paving the parking lot.

I remember sitting on a panel moderated by Dick Morley, the inventor of the programmable logic controller and industry guru, at an AMR Conference years ago. As Dick started questioning the panel comprised of technologists from most of the major automation companies, he asked, "Who are your competitors?" We all looked around at each other with confused looks on our faces as if to say that we were each other's competitors. He immediately added, "Don't look at each other. You are competing with the storm windows in the building in northern Minnesota." At that time I was a bit taken aback by his statement, but he was exactly correct. Automation and IT investments are competing against all other possible capital expenditure. Whichever investments offer the best economic returns get the budget. The ones that do not, do not! It is fairly easy to measure the economic impact of storm windows on a building in Minnesota. Even if an automation or IT project provides great incremental value, if the value of automation and IT capital investments is not measured, over time, management will redirect capital funds to other projects that are measureable.

Even if the ROI of an automation or IT project is measured after project completion, ROI tends to be a negative concept. Typically, ROI is measured until the investment is completely recaptured. That is, once 100% of the project cost is returned in the form of incremental value, most companies stop counting. This means that the return to investment is always either negative or zero. Once zero is hit, we stop counting. Automation and IT investments continue to provide value even after 100% ROI is achieved. That incremental value must be made visible to management in order to encourage additional investments in these value-generating solutions. Perhaps, a better way to measure the value from automation and IT is by cash flow since the cash flow tends to continue for years into the future.

These issues and barriers associated with capital budgeting processes are easily overcome, but industrial management must understand that they exist in order to fix them. The first change that is critical to success is to measure the project teams on the value generated by the solution as well as on time, on budget project delivery. This will encourage the project teams to make better decisions for the business. In order to accomplish this though, a system must be put in place that effectively allows for the measurement of the value generated by each capital project. This can be effectively accomplished by installing the real-time accounting systems previously presented as part of a real-time enterprise control system. This can cause an interesting "chicken and egg" scenario when transitioning to real-time enterprise control. If management understands the value of measuring results, which every good management team does, they can implement the real-time accounting system across each site first and then implement each of the other aspects of the real-time enterprise control system incrementally. In this way, the value of each component of the overall system will be discernible in accounting terms that can be audited by the accountants. Furthermore, the presentation provided by the real-time accounting system enables ROI calculations, but more importantly, can inherently present a "cash flow" view for each investment since the resulting data are real time based, can be viewed right down to the asset level, and typically come with a trending mechanism over time. With a slight shift in perspective on the mechanics of capital budget processes and with the functionality provided by real-time enterprise control systems, the capital budget process can easily be made to transition from a barrier to success to an enabler of success.

TRADITIONAL ACCOUNTING PRACTICES

The accounting systems and approaches used in industrial companies today were based on necessary compromises caused by the increasing speed and complexity of business resulting from industrialization. As new manufacturing machines were developed enabling faster and faster production of goods, piecemeal accounting was no longer viable. Industrial companies moved to monthly accounting systems as a compromise because it was the only practical approach available. Entire business infrastructures and value chains adjusted to monthly accounting. It became—and still is—the norm in industry.

Monthly accounting systems have been used to report both financial results and provide a level of business management decision support information. As long as the key variables of the business of industry were fairly stable for over a month at a time, the information provided by the monthly accounting systems was typically sufficient for making business decisions associated with those variables. As the speed of business increased, especially over the past decade, many business variables started to experience variations in submonthly and eventually even subdaily time frames. As this took place, the decision support value of monthly accounting data was significantly reduced. If, for example, the cost per unit of electricity changes every 15 minutes, monthly measurements of electricity cost provide little or no decision-making value. The monthly accounting systems might still be sufficient for reporting overall financial results, but new, higher-resolution approaches to cost accounting are required to provide effective and timely business decision support.

Although with today's technologies, real-time accounting systems that operate underneath the existing financial reporting systems are not highly difficult or complex to develop, few industrial companies are even considering moving in this direction. Perhaps, the reason is that we have become so locked into the traditional approach to accounting that it has almost become culturally impossible to consider an alternative approach. Coming out of the Industrial Revolution the complexity associated with managing industrial businesses increased almost exponentially. To make industrialization feasible industry had to quickly learn how to manage these businesses in as simple and practical a manner as possible. The financial accounting systems and monthly financial accounting approaches were developed to meet these needs. Colleges and universities took up the challenge of supporting these approaches and developed accounting and business management programs to teach students how to be effective in dealing with the monthly financial accounting systems. Once a set of students had learned the "right" way to do accounting, and actually had been awarded degrees in the traditional monthly accounting systems, it became close to impossible for them, as a group, to consider the real-time accounting—or any other—alternative. Today, monthly financial accounting systems are locked into the business management infrastructure and business culture of industrial companies.

Over the past decade, as monthly financial accounting systems have proven to be inadequate for the job of real-time business decision support, it has been very difficult to get management of industrial companies to consider alternative approaches that align with today's business dynamics. The good news is that monthly financial reporting is still very much required for industrial companies. Therefore, the existing monthly accounting systems, normally part of an Enterprise Resource Planning (ERP) software system, still perform an important function. They are absolutely necessary—just not sufficient. Utilizing the installed real-time automation technology in conjunction with ERP software to develop a comprehensive bottom-to-top financial system is the key to success (Figure 6.3). This can be effectively accomplished by building the real-time accounting models and operating them in the real-time automation system environment, historizing the real-time accounting data through process historians and uploading the historized accounting data into the financial reporting systems in the ERP software. The real-time decision support must be provided across the organization according to the business decision rights of the organization and can either be sourced by the real-time

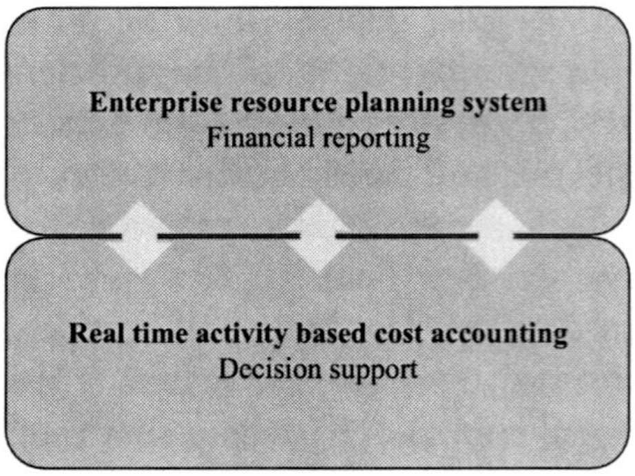

Figure 6.3. Industrial financial accounting architecture. With the transition to real-time business variability a two-tier financial accounting architecture that covers both real-time financial decision support and financial reporting functionality is required to support effective profitability control.

accounting models or the ERP software or both, depending on the required data resolution for the decisions being made.

Although the concepts associated with a real-time financial accounting and management system are truly just an extension of current ERP system approaches, the performance improvements that can result can be very significant. Cost accounting systems have not traditionally been considered to be integral to either performance management or continuous improvement programs, but they must be used for the desired results to be realized. Financial and business managers must expand their perspectives on what needs to be accomplished by the financial systems. However, the results created by the expanded systems will certainly reward the business.

COMMON ENGINEERING PERSPECTIVES

Perhaps, the most valuable positions in any industrial company are in engineering. Engineers were relied upon during the early segments of the Industrial Revolution to design the manufacturing and production processes to make them efficient, safe, and profitable. Engineers have had to take the lead in driving performance improvements during all phases of industrial history. Industrial management has consistently leaned heavily on engineering to ensure that the operations are performed as intended. The history of the Industrial Revolution demonstrates how effectively they performed their functions.

To a very large extent engineering has taken a leadership role in making sure that frontline laborers were performing their duties as desired. This has meant that as the operators managed the ongoing control and safety of the process, the engineers had to ensure that the objectives would be effectively met by the operators. Traditionally, the process operators were considered unskilled positions that were typically filled by uneducated laborers. As a result, management and engineering often believed that they could not trust the operators to make the best decisions or even have reasonable work habits. In a very real sense, this often meant that the engineers had to design the processes and process control systems to protect the plants from the operators. This has resulted in a cultural schism between engineering and operators.

Over the past three decades, as more and more advanced technologies were introduced into industrial automation to help make plants and processes more efficient the schism grew. Engineers invested considerable time, talent, and technology to design and develop advanced control and optimization strategies to make the plants more efficient and profitable. Operators, who did not understand or trust these advanced strategies would often wait for the engineers to leave the control room so that they could turn the strategies off and manage the processes manually. This was a constant source of frustration to the engineers. An extreme lack of trust developed between operators and engineers. Today the schism between engineering and operations is typically built into the culture of industrial organizations and is very difficult for both sides to breach.

As strategies evolve to provide better control of real-time enterprises, the schism between operators and engineering is presenting a considerable barrier to success. One of the basic premises for effective control of real-time enterprises is that the role of operators must evolve from that of a laborer into a performance manager. This can only happen if engineers enable the operators to make this leap through new technologies, applications, empowerment, and support. This is a huge challenge for engineers who do not believe that most operators are up to the task and are culturally tied to traditional ways and perspectives.

Different companies are taking different approaches to deal with this barrier. Some industrial companies, especially in the developing areas of the world, are starting to hire operators who are also trained and degreed engineers. Certainly, the process and control engineers tend to put more trust into the emerging operator–engineers than they did into the traditional process operators. Although the movement toward operator–engineers seems like a logical approach, there are engineering shortages in some geographies, and especially in the traditional industrial areas. In these areas, moving toward an operator–engineer combination is not yet practical.

A second approach to deal with this is to start by transitioning the most senior operators for a contained portion of the overall process. Once these senior operators have proven proficiency for a small portion of the operation, the scope of their responsibility domain can be increased until they eventually take responsibility of entire plants and perhaps even multiple plants across an industrial enterprise. As the senior operators realize success in controlling the real-time enterprise, less-experienced operators can apprentice under them. This is very similar to the traditional approaches to bringing new operators into the business. With the advanced training tools available today, such as simulators and virtual reality systems, this type of apprentice-based approach can be accomplished much more efficiently than it could have been done previously.

Regardless of which approach is followed, every industrial organization must come to terms with the transition of operators from laborers to performance managers for the move to enterprise control to be successful. Experience has shown that academic education is not necessarily the best predictor of operator success in making this transition. Experience and experiential training appear to be better predictors. In any case, the engineering staffs of industrial companies must undergo a cultural perspective change associated with the plant and enterprise operators or the transition to enterprise control will not be effective.

A second issue associated with engineering that is a concern with respect to the move toward controlling the real-time enterprise is that the discipline of control engineering has been in decline for two decades. As recently as the 1970s, good control engineers were considered to be critical to the success of manufacturing and production operations. Today it has become difficult to identify any control engineers across industry. Perhaps, one reason for this is that with the advancement of automation and control software technology, basic regulatory control has become simpler to implement, tune, and maintain. Many industrial companies that used to have engineers dedicated to control have determined that they no longer need this dedicated resource. The great control engineers of the past, such as Greg Shinskey, Carroll Ryskamp, Edgar Bristol, John Ziegler, Nathaniel Nichols, and Greg McMillan, have retired and their replacements are becoming more and more difficult to find. Those control engineers who are still in the profession do not seem to be in great demand. Basically, the field of control engineering is in a slow decline and has been declining for decades.

This is a huge problem for industry. Many of the challenges industry is facing are control and control theory challenges. These challenges may be in areas beyond the basic regulatory control in plants and factories, but they are control problems nonetheless. Controlling the real-time enterprise involves applying control theory to processes, safety, assets, environmental issues, product quality, and profitability among other areas. Doing this effectively requires control engineers to be flexible enough to apply control theory in key areas never considered previously.

Looking back at the beginning of the Industrial Revolution before the development of standard automatic control approaches, such as proportional, integral, and derivative (PID) control, engineers often used operators to close the control loops manually. This was done by empowering the operators with control measurement information as displayed on gauges and allowing them to manipulate hand valves to control the variable of interest. Since standard algorithms and approaches for controlling business variables do not yet exist, similar manual control of business variables may be the initial approach. By providing operators with performance dashboards or scorecards with real-time performance information, the operators can view the values of interest and change set points or working recipe variables to see if they can get the variables to move in the desired direction. This is essentially a manual feedback business control loop. It is interesting to note that learning sciences has determined that humans tend to commonly utilize feedback control as the standard learning approach. This was demonstrated at the Massachusetts Institute of Technology artificial intelligence labs back in the 1970s. We often refer to certain types of learning as "trial and error" approaches. Trial and error is a classic feedback control mechanism. The reason this is important is that it points to the use of human operators as an effective component of a feedback control system. Today we can use operators as the basis for effective manual control of business processes over the automated process control systems in place to control the efficiency of the manufacturing and production processes.

In any case, it has become clear that one of the key enablers for systems to control the real-time enterprise is control engineering. Many industry observers believe that the golden age of the control engineer was in the past. In fact, for the movement to control the success of real-time enterprise, the golden age of the control engineer is actually coming.

ORGANIZATIONAL SILOS

For years comments have been made across industry with respect to the "islands of automation" or the "islands of information" or the "islands of technology" that exist throughout industry making it difficult to pull industrial operations together into a single, unified operation. The technology seems to have been used as the reason for the lack of unified organizational approach. However, as the technology has become more open and standard enabling greater degrees of interoperability than was previously practical, the islands persist. Perhaps, the technology has been used as an excuse for what may not be a technology problem at all—"islands of organizations."

Islands of organizations or organizational silos—whichever phrase you prefer—are presenting a huge barrier to controlling the real-time enterprise as a unified business.

These organizational silos were actually developed early in the Industrial Revolution for very good and practical reasons. With industrialization the complexity of both the manufacturing processes and the business of industry increased to an unprecedented extent. New, specialized skills and talents were required to deal with the complexity. Each of the advancements resulted in new skill requirements that presented daunting challenges in their own right. Almost overnight specialized talent was required in areas such as the operation of the machinery, the maintenance of the machinery, engineering of design and performance improvements, accounting, human resource management, business management, supplier management, client management, safety management, and the like. Practically, meeting these needs necessitates that companies developed excellent talent in each of these specific functional areas. As this took place, organizational silos formed. Each silo was comprised of similarly skilled and trained talent who had their specific knowledge base and even a specific vernacular associated with the needs of the silo. Over time as the complexity of each silo increased it became very difficult for the people in one functional area to understand what those in the other functional areas actually did. It even became difficult for good conversations to take place between the people in the different silos because they used words unique to their function that the people in the other silos had difficulty interpreting. After some time the silos tended to become more and more isolated from each other. When the underlying automation and IT reached the point at which inter-silo interoperations were well supported, the people in the silos had difficulty taking advantage of the technology.

There are a number of examples of key areas of organizational silo behaviors that must be overcome in order to facilitate controlling real-time enterprises. For example, business management and operational management in many industrial organizations have great difficulty converging. Business managers are focused in the world of accounts receivable and payable, shareholder management, financial reporting, human resource management, strategic relationship management, and similar business functions, while operations managers are focused on throughput, efficiency, production schedules, safety, environmental management, and similar operational functions. Obviously, the performance of operational managers significantly impacts the business management results and the decisions of business managers impact operational efficiency. Traditionally, these two management groups have had great difficulty reconciling between each other. Even the technology bases supporting these two groups have been fairly independent with business systems supporting business management and automation systems supporting operational management. Over the past few decades there has been considerable discussion on how these two classes of systems should interact and connect, but in many industrial operations, effective interoperation is yet to be achieved. This is due more to a lack of understanding on how each group can support the other than on limitations of the technology. Standardization activities such as the S95 standard from the International Society of Automation (ISA) are beginning to provide stronger convergence discussions.

A second example is closely related, that is, the relationship between the automation technology and IT organizations in industrial companies. Due to the traditional business management and operations management silos and the independent technology development paths, the automation and IT teams in many industrial operations act very independently. Today, both business systems and automation systems are based on many of the same digital computer, information management, and communication technologies, and there is pressure from management to provide solutions across business and operations management that can drive profitability improvements, but automation and IT teams typically operate independently. There has been some movement over the past decade in a number of industrial companies to converge these two organizational silos into single technology solutions teams. This can be quite painful, but over the long run can help to drive a much stronger unified approach. IT professionals tend to have strong knowledge and skills in the areas of technology infrastructure, security, communications, and information management while automation engineers tend to have strong skills in industrial problem solving. Together, these two groups can offer significantly more value to their organizations than they cannot separately.

The fundamental differences between the perspectives and culture of IT and automation teams are difficult to overstate. Real-time control talent in the automation engineering group is typically focused on using technology to drive operational efficiency. These engineers often have distaste for business due to their lack of understanding of business processes and how they fit in. They tend to focus on the natural sciences and business is often seen to be an unnatural science, in that it is human designed. The automation engineers often are lacking in business and IT expertise. It is very important for industrial companies to identify this barrier to success and work to overcome it in order to drive toward controlling the real-time enterprise. The computer systems in plants were designed, configured, and programmed to capitalize on engineering know-how and abstract the details of the information and computer technology from the engineers. IT teams, on the other hand, tend to have expertise in fundamental computer science, infrastructure, and information technologies. They tend to be less experienced in real-time computing and real-time control. In fact, the IT talent and the automation engineering talent typically are quite complementary.

Unfortunately, due to the history at most industrial firms, these two groups do not tend to trust each other. This may be due to the close alignment of mission between IT and automation coupled with the pressure to pull the technological platforms associate with each team together into a unified platform. Each group may be concerned that the other will take over its domain.

Management is looking for value-adding solutions to daunting business challenges. If the IT and automation teams would assess their competencies, they may find fairly good complementary alignment that could be very effectively utilized to meet the daunting business challenges. Some industrial companies, such as Eli Lilly, have made great strides in pulling these two talent pools together and are starting to experience good results.

A third example of organizational silos proving a significant barrier to successfully controlling the real-time enterprise can be found in many accounting and engineering departments. Accountants and engineers tend not to work closely together and, once again, tend to distrust each other. The move toward real-time accounting and ERP will require that accounting and engineering work very closely together. Engineering has access to the real-time sensor-base of the industrial operations and understands real-time computing while accounting understands the applicable basic accounting approaches, rules, and generally accepted accounting practices (GAAP) used in the plant. If these two organizational silos would work together toward a common goal, they would learn that each possesses critical puzzle pieces in the move toward a comprehensive and validated real-time accounting approach.

One final set of organizational silos that provides a barrier in the move toward controlling the real-time enterprise is represented by the lack of traditional cooperation between plant operations and maintenance. It is truly astounding that two organizations working on the same set of assets tend to work independently, but in many plant organizations, operations and maintenance actually report through totally different paths to the plant managers. Visits to industrial plants often reveal that operations and maintenance departments do not cooperate, are at odds with each other, and do not even like each other. Perhaps, the reason is based on the primary performance measures of these two groups. The operation team is typically measured on the production throughput of the operation while maintenance is often measured on the availability of the production assets to operations. These two basic performance measures are inverse functions. Typically, increasing throughput involves reducing asset availability and increasing asset availability involves reducing throughput. It should not be surprising that these two teams are at odds. New management approaches with new performance measures are required to encourage the cooperation between operations and maintenance that will drive increased profitability for the operation. The new management structures required, encouraging the appropriate level of cooperation between operations and maintenance, will be presented in the next chapter.

There are a number of additional organizational silos that present a difficult barrier to the transition to the real-time enterprise. The ones presented herein are among the most critical, but industrial management must keep a watchful eye on organizational behaviors to drive to a unified, performance-based approach. Technology presents some barriers to success, but the barriers based on culture and organizational behavior are much more difficult to overcome.

BUSINESS AND PRODUCTION PROCESS LIMITATION

One final set of barriers to success needs to be understood when trying to move toward controlling the real-time industrial enterprise. Those barriers are provided by business and production processes that are not agile enough to meet the timing requirements of the emerging real-time business environment. The focus of this book is on the real-time control of both production and business processes. However, it is important to realize that control can only make the processes work to their designed intent and no better. If the time constant of a production process is longer than what is necessary to effectively control the real-time profitability components of the operation, the control system cannot compensate or overcome the design limitation. For example, if a company is shipping crude oil through a 1,000-mile-long pipeline and it takes five days for the crude to pass through the pipeline to the refineries. The five-day time constant can provide a significant barrier when trying to maximize the profitability of the business. The price of the crude may change well over 100 times during its trip through the pipeline and there is nothing any control system can do to compensate for the process constraint.

Many industrial companies may need to reconsider some process design decisions they have made in the past. The company using the pipeline to move the crude to the refineries may want to consider actually building a refinery closer to the crude source to reduce the transport process time and provide a greater degree of agility with respect to controlling profitability.

Throughout the 1980s, industrial companies discussed the concept of just-in-time (JIT) manufacturing to reduce the storage capacities at the front and back ends of the manufacturing operations in order to realize storage cost savings. The concept of JIT manufacturing may make a strong resurgence—not just to reduce storage cost—but to increase the agility of the overall production processes to be able to match the time constants of the real-time business environment. Developing storage capacity in various locations across a value chain may make the production logistics at each point a bit easier to deal with, but this is at the cost of agility. Business and operational agility is starting to increase in impact as real-time decisions are tied more closely to the profitability of industrial companies.

A similar consideration and evaluation of business processes must be made to ensure that the existing business processes are not providing constraints to profitability improvements. As was previously discussed, many of the business processes existing in industrial companies were developed as pragmatic compromises as the Industrial Revolution occurred. Those compromises served industry well for many decades. With the move toward real-time business variability, some of the previously made compromises are working against improved levels of profitability and must be altered. For example, it was pragmatic for power companies to communicate to the plants through an electric bill provided on a monthly basis. Since the price of electricity did not change often, or much during monthly periods, this compromise made perfect sense. But today, with the price of electricity changing multiple times daily, the traditional monthly business process is inadequate. The process must be changed to provide the degree of agility necessary to control the profitability of the business.

As industrial companies grapple with real-time business variables, many production and business processes will need to be reevaluated along with the process and business control systems. The processes will only be able to realize the potential profitability improvements through real-time process and business control if the processes are agile and fast enough to respond to real-time profit-based decisions.

There are a number of significant barriers that can prevent the move toward controlling real-time industrial enterprises. Each of the barriers is significant in its own right, but taken together, they can appear to be almost overwhelming. Identifying and understanding each barrier and working to resolve them one at a time may present the only viable strategy. Unfortunately, most of the barriers are cultural and are tied to ingrained organizational behaviors. Cultural-based barriers can be much more challenging to address than technological barriers. It will take a focused effort for industrial companies to move forward toward the real-time enterprise. Any plan for moving toward controlling the real-time enterprise must involve deliberately identifying and overcoming each of these barriers to success. It is certainly a daunting challenge, but the results in terms of increased profitability are certainly worth the effort.

7

Practical Applications of Real-Time Enterprise Control

Developing a Real Time Enterprise Control (RTEC) program is truly a daunting challenge. Trying to develop an overall plan and strategy to implement an RTEC system can certainly be overwhelming. It can be very helpful to establish a clear starting point. Selecting an initial focus that aligns with the organizational and cultural strengths of the company and provides a reasonably well-bounded solution space can be a critical success factor in effectively executing the program.

Over the past couple of decades, a number of industrial companies have embarked on a similar transitional exercise under the heading of "operational excellence" systems. Like the move toward RTEC, the scope and breath of an operational excellence initiative can tend to overwhelm even the strongest industrial organizations. Traditionally, some of the more effective approaches to developing and executing an operational excellence strategy seem to have realized a level of success because they initially focused on specific areas of success, such as asset management, operations management, process control, or safety. By initially focusing on a single area of improvement and making key gains in that focused area, the program teams found that they could realize initial success and use that success to leverage the improvements to other areas. The more the initial focus area aligned with the organizational strategy and culture, the more success the organizations seem to have realized. A similar strategy may help structure the transition to RTEC.

In fact, many of the business drivers associated with operational excellence programs are identical to those associated with the move toward controlling the real-time enterprise. The focus on controlling the real-time enterprise is fundamentally different from a traditional operational excellence focus in that it goes beyond strict operational improvements and also incorporates business and profitability improvements. Many of the specific improvements required for good operational excellence are also prerequisite to effectively implement RTEC. Therefore, in developing an effective strategy and plan for RTEC, following the same basic operational excellence focus areas with a larger business and profitability context makes perfect sense.

Experience has demonstrated that there are four specific focus areas that can be effectively utilized to get an RTEC program going. The four areas can be thought of as different paths to solving the same problem since they tend to overlap fairly significantly. They are:

- Control Performance Excellence
- Asset Performance Excellence
- Human Performance Excellence
- Safety and Environmental Performance Excellence

Control Performance Excellence involves effectively applying control theory to control both the efficiency of the operation and profitability of the business of industrial plants and enterprises. Asset Performance Excellence involves maximizing the financial value provided to the business from each major asset and asset set across an industrial enterprise in an ongoing manner. Human Performance Excellence involves empowering the personnel in the operation and across the enterprise to make better decisions that drive better operational and business results, and Safety and Environmental Performance Excellence involves dealing with the primary variable real-time constraints on profitability, which tend to be safety and environmental limits.

Many industrial companies have already aligned themselves around one or more of these strategies. It is often helpful in getting an RTEC program started so that the organization builds on its traditional strength and focus by starting with one of these four strategies that aligns best with its traditional approaches. This can help the organization to build on any momentum of the past operational excellence initiatives and gain organizational acceptance. These four strategies overlap significantly, as will be better realized as the specifics of each of these strategies are reviewed. It does not really matter which strategy a company selects as its launching point into an RTEC program. Starting with a single strategic focus will help the organization to better understand and execute the overall program. It will remove some of the mystery associated with a broad concept such as RTEC and make it seem more real, bounded, and doable.

The ultimate objective of an RTEC program is to implement the infrastructure and functions that drive real-time profitability improvements. Each of these four strategies is a different approach to accomplish this goal and is

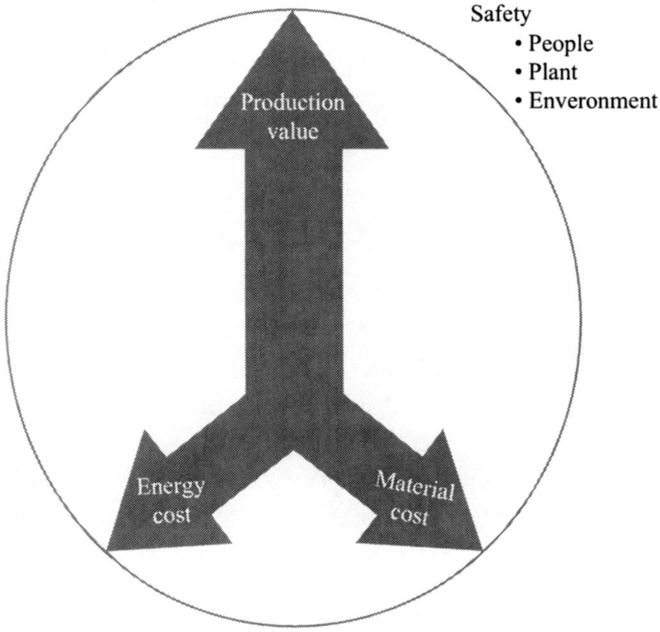

Figure 7.1. Constrained industrial real-time profitability model. The constrained real-time profitability model applies throughout industrial enterprises—across multiple asset sets—not merely at the single asset level.

based on the same principles associated with the real-time profitability model previously discussed (Figure 7.1). Prior to getting into the details of each of the strategies, it may help to review the real-time profitability model and concepts as presented in the model. The three components of industrial business profitability that tend to experience real-time variation are the production value of the products being produced, the cost of the energy required to produce them, and the cost of the material required to make the products. In recent years, all three of these variables have started to fluctuate in time frames considerably less than a day, making the control of the profitability of industrial operations especially challenging.

There are a number of constraints to driving real-time profitability. These include equipment constraints, supply constraints, and the like. However, two constraint categories present the most dynamic limitations to industrial processes: safety and environmental impact. Since both safety and environmental issues can change with production conditions, both of these constraint functions tend to experience real-time or near-real-time variation. This means that the real-time profitability model of Figure 7.1 is dynamic across all five variables: production value, energy cost, material cost, safety, and environmental impact. Each of the four RTEC strategies is designed to positively impact real-time profitability in a safe and environmentally sound manner. Executing all four of these strategies, whether in series or parallel, provides a fairly comprehensive RTEC program approach.

CONTROL PERFORMANCE EXCELLENCE

Control theory has been applied to the effective automation of manufacturing operations for decades with the primary objective being improving the efficiency of the industrial operation. The variables associated with the efficiency of the operation, such as flows, levels, temperatures, pressures, compositions, and speeds have been measured and controlled in real time for decades. The effective application of control theory was absolutely essential for operating many industrial plants at any reasonable level of efficiency and safety while minimizing the negative environmental impact from these operations. Advancements in the science of process control have been continual for decades. Today, process plants can be controlled at efficiency levels never previously possible through the effective application of these advancements in control.

Traditionally, controlling the efficiency of industrial plants directly translated into improved profits for the industrial business. Today the business case is not quite as simple as it had been. Critical business variables, such as energy cost and material costs, are changing at very frequent rates, perhaps every 15 minutes or faster. With this situation, two plants may consume the same amount of energy over the same time period and one may pay a significantly greater amount for the energy it consumed because it consumed the energy during higher price time periods. This can also be true for material costs and even production value. Today, controlling efficiency is just not enough. For maximum business results industrial enterprises must also control real-time profitability.

Control performance excellence involves simultaneously controlling plant efficiency and enterprise profitability—both in real time. It is critically important to realize that if the basic process control systems are not working to potential and controlling the efficiency of the plants, there is little or no hope of implementing a real-time profitability control system. Perhaps, the best way to think about this is as a simplified cascade control strategy (Figure 7.2). The primary loop of this strategy is controlling the real-time profitability of the business and the secondary loop involves controlling the efficiency of the operation. As with any cascade control strategy, if the secondary loop is not well controlled and tuned, the primary loop will not be effective. Therefore, the first step in implementing a control performance excellence approach

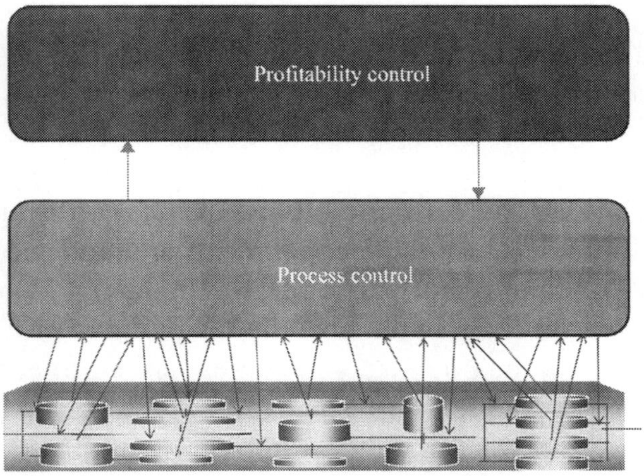

Figure 7.2. Profitability control cascaded to efficiency control. As with all cascade control strategies, the primary controller (profitability control) cannot operate effectively if the secondary control loop (efficiency control) does not operate effectively. Therefore, the first step to a profitability control strategy is ensuring that the process control system is complete and working correctly.

across an industrial enterprise is making sure that the basic regulatory and advance control systems in each plant are in place and working well. In this way, the efficiency of each plant will be well controlled.

There have been numerous books written on regulatory process control systems and advanced process control systems. This information will not be repeated herein. It is absolutely imperative that the plant control systems operate well prior to moving forward with a real-time profitability control system. The reason this is emphasized is that over the past few decades, as new ideas and technologies have been introduced into industrial automation, users often jump over the basic technologies to the new technologies. In the case of control performance excellence, this would be a costly mistake.

Once each plant's process control system is installed and it operates effectively, the operation can move forward with real-time business control. The recommended approach for the initial implementation of a real-time business control strategy is based on the approach utilized early during the Industrial Revolution for basic plant process control systems—manual control. In the early days of industrialization, process operators were stationed at manual valves and had access to gauges that display the value of the critical process variables. The operators acted as the controllers by manipulating the value of the variables of interest by adjusting the values until the values of the variables reached the desired values, or the set point (Figure 7.3). These manual control systems worked quite well. An exactly same approach can be very effectively applied to the control of the real-time profitability variables.

Most of the processes for implementing a manual real-time business control approach were previously covered. They involve determining how to measure and prioritize the real-time business variables through a Vollmann Decomposition process. Once the measurements are modeled and prioritized the appropriate Dynamic Performance Measurement (DPM) dashboards can be developed for each person in the operation with direct profitability impact. The dashboards should be bounded and contextualized to each person's specific domain of responsibility. In this manner, as each person performs his or her specific duties he or she can monitor his or her dashboard display to see if his or her action had a positive or negative impact on his or her critical variables of responsibility. Over time, each person will learn how his or her actions can drive improved performance and will use this knowledge and the real-time feedback to control the key real-time business variables. This is very similar to early operators using gauges as empowerment dashboards to control the process variables of production operations.

With experience, industrial businesses will likely want to move to automatic business control. Unfortunately, this does not appear to be as straightforward as the general approaches used to control production variables. These general

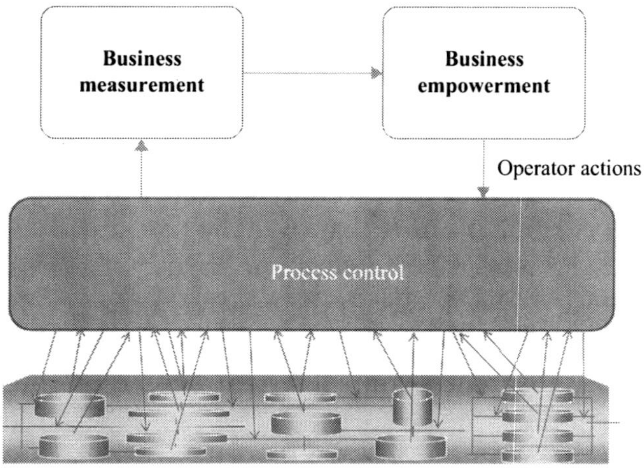

Figure 7.3. Real-time business control through empowerment. Early during the Industrial Revolution most process control was done manually with human operators making the control decisions. As business control strategies are developed a similar trend is expected with manual business control through real-time empowerment.

approaches, such as proportional, integral, and derivative (PID) control algorithms, depend on a natural period of the control loop. Physical and chemical processes tend to have fixed natural periods, but business control loops do not. This means that the most likely approach to closed-loop automatic business control will be the development of process-specific control approaches most likely combining algorithmic and heuristic approaches with modeling. One of the more promising approaches being tried is workflow analysis. This analysis examines the processes in terms of workflow and sets timings accordingly. This may make it possible to have process sensors and scheduling tools speaking the same language. It may take some time for industry to get there, but the improvements will likely be worth the effort. Until then, industrial companies can deploy traditional improvement strategies within their plants that may drive the values in the desired direction, such as advanced process control, multivariable predictive control, advanced maintenance approaches, quality improvement programs, or other model-based control strategies (Figure 7.4). The advantage of executing these strategies within an overall Control Performance Excellence approach is that the installed DPMs will measure the business improvement realized from each. Companies can then focus on those that drive improvement and move away from those that do

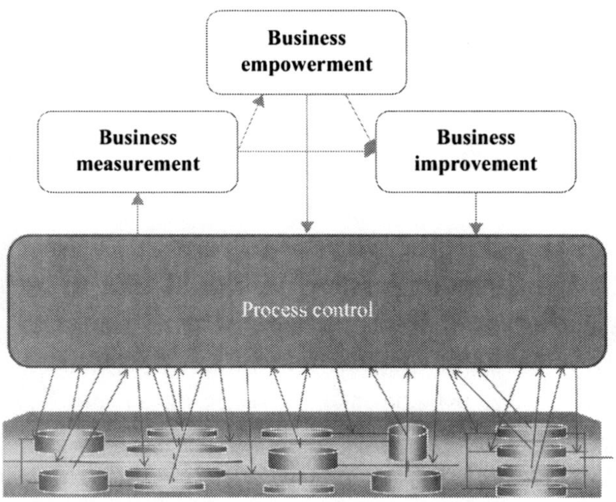

Figure 7.4. Real-time business control. As with process control loops, business control loops have four basic components: the business process to be controlled, business measurement, business empowerment, and business improvement. With the real-time business measurements installed, the value of any improvement will be automatically measured from a business perspective.

not. This will enable a concentration of resources and effort on only those tools that provide the best improvements.

Companies that have already started to execute a control-based approach to achieve operational excellence would be well advised to extend that focus to Control Performance Excellence. Once the basic controls of the production process have been effectively implemented, the manual business control strategies can be developed. Once these are in place, the specific improvement strategies can be implemented. The net result will be a very major step toward an overall RTEC strategy. Industrial companies can move plant by plant, or take an overall enterprise-wide approach depending on the availability of implementation resources. Once a Control Performance Excellence strategy has been implemented, the company can move to the next of the four strategies that matches its cultural initiatives and its way to a fully implemented RTEC system.

ASSET PERFORMANCE EXCELLENCE

Today when the phrase "asset management" is used in industrial operations it typically equates with the maintenance of plant assets. Management's perspective on asset management is quite different. To industrial management the phrase typically implies the financial management of all capital assets across the business and operational enterprise. Certainly, keeping the plant equipment in operational equipment through an effective maintenance program impacts the financial performance of the assets. Therefore, in this respect, the two perspectives on what asset management means do overlap. However, maintenance is not the only activity that impacts the financial value generated by industrial assets. Extending the plant-level context of asset management to better align with management perspective would lead to better communication and strategic alignment from management to the plant floor. Since the plant floor semantic for asset management is fairly well fixed, the phrase asset performance management will be utilized to management's perspectives down to the plant-floor level.

The application of technology to the control and operation of industrial processes has been a primary concern across industry for decades. For this reason, the science of process control and operations is quite advanced. The technological aspects of plant-level asset management were not as aggressively pursued until the last decade or two. Although asset management technology has been advancing rapidly, it is not quite as advanced as the technology associated with process control and operations. As a result, a number of industrial companies have made progressing in the application of advanced asset management technology a main initiative for their industrial operations.

These industrial organizations tend to have a strong asset management culture. Initiating a move toward controlling the real-time enterprise from an asset management perspective may be the easiest and most direct route for these organizations.

Within the asset performance management context, the two plant organizational teams that have the most immediate impact on the performance of the industrial assets from day to day are the operations and maintenance teams. The operators drive the assets while the maintenance team works to keep the assets available to operations. It should seem obvious that the closer these two teams collaborate on driving the performance of the assets, the better should be the potential for asset performance improvement. Unfortunately, in most industrial plants, the operators and maintenance teams do not tend to work very well together. In fact, in many plants, they do not even like each other. The reason is that the traditional performance measurement priorities for operations and maintenance tend to be in conflict with each other (Figure 7.5).

Operators are typically measured based on maximizing the utilization of the assets that lead to maximum plant production throughput, while maintenance tends to be measured based on the availability of the assets. Increasing utilization typically requires a reduction in availability. Increasing availability often requires a reduction in utilization. In essence, availability and utilization are inverse functions (Figure 7.6). Measuring any two organizational teams in

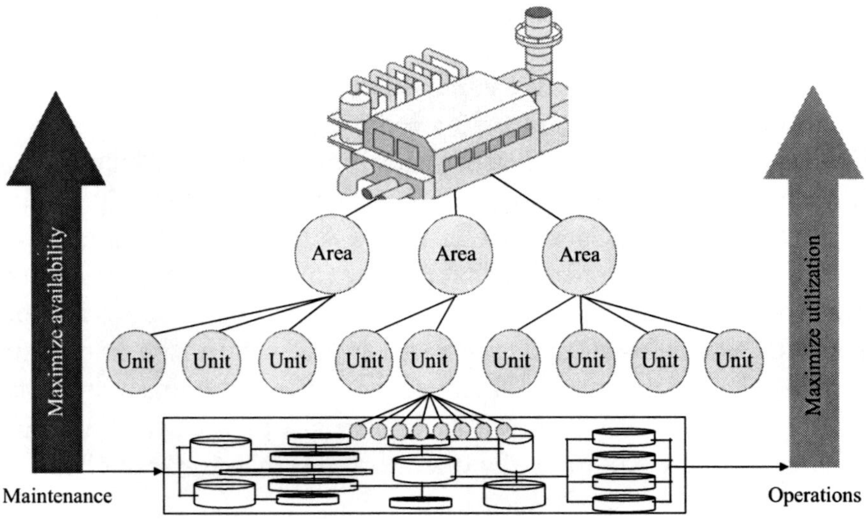

Figure 7.5. Traditional plant conflict. It is no secret that the maintenance and operations teams of many industrial operations tend to be in conflict with each other. The genesis of the conflict is conflicting performance measures.

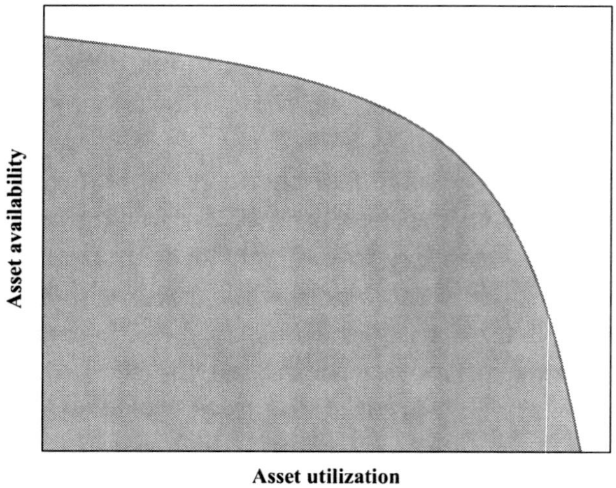

Figure 7.6. Industrial conflict—inverse function. Traditional performance measurements, such as measuring maintenance based on asset availability and operations based on asset utilization, often resulted in organizational conflicts because the measures tend to be inverse functions. Increasing availability may require reduction in utilization while increasing utilization might negatively impact availability.

an inverse manner will drive conflict between the teams. Is it any wonder that operations and maintenance do not collaborate—or even cooperate?

The starting point for asset performance management is getting the primary plant teams responsible for day-to-day performance of the plant assets to collaborate. Perhaps, industry could learn from the experience of a highly performance driven organization in which the operators and maintenance teams collaborate at very high levels with very good results. Hendrick Motor Sports is a performance-based organization that very successfully competes in NASCAR (National Association for Stock Car Auto Racing) racing. Interviews with Hendrick personnel revealed that the key to their collaboration is that the operators (drivers) and maintenance team (pit crew) have a common primary performance measure—winning the race. In car racing, the end-in-mind is very clear and easily measured. Whichever car crosses the finish line first wins. What is the "win" when it comes to industrial operations? What common performance measure will drive the operators and maintenance people to collaborate to drive improved performance?

Taking the definition of asset performance management as previously presented provides a clear definition for *winning* in industrial operations. The win in industrial operations is driving the most financial value for the operation from each of the industrial assets. Since the concept of DPMs presented earlier

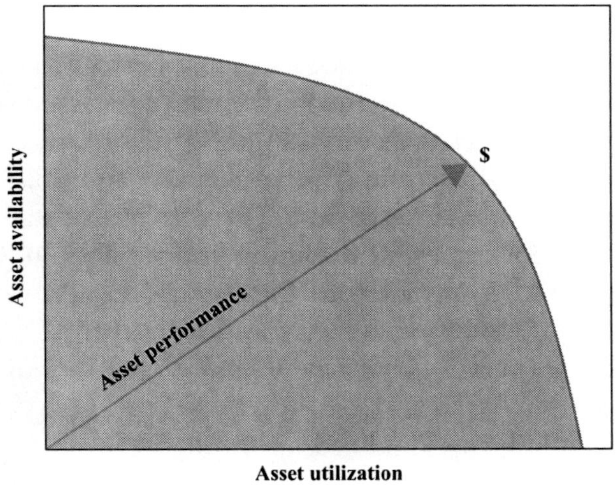

Figure 7.7. Asset performance management. Resolving organizational conflict often depends on aligning the performance measures between departments. By making asset performance (the business value driven from an asset or asset set) the primary measure of both operations and maintenance, the two teams tend to cooperate and collaborate rather than fight.

provides the ability to measure the financial value being generated down to the asset level, the prioritized DPMs provide the basis for operations and maintenance collaboration (Figure 7.7).

Transitioning to asset performance management for industrial companies with a strong asset management culture can be systematically accomplished by following a bottom-up four-step process for each asset in the operation and then combining the assets together to create an overall asset performance environment for each plant.

The first step is to install the DPMs right down at the industrial asset level. The real-time accounting models will serve to measure the financial value for each asset. The DPM models can typically be installed in the automation systems already operating in the plants. The DPMs should be historized in the plant historian. The historized data will provide a time context to the asset performance. This time perspective is important because the trends over time will enable the asset degradation with use to be discerned. Industrial companies may want to explore advanced analysis techniques, such as prognostics, to analyze the optimal performance point for maintenance procedures with respect to the value from the assets over time. Since industrial assets are typically interrelated, analyzing asset sets may be required for optimal results.

The second step is to train the operations and maintenance personnel with respect to their new performance focus and performance measures. The prioritized DPMs should be the true performance measures for both operations and maintenance. Having common performance measures defines the win and encourages collaboration. Asset availability and utilization should be relegated to secondary measures at best. These measures should be primarily used for analysis of root causes for declining performance in the performance measures. People tend to perform to their measures. Providing operations and maintenance with common measures prioritized to the financial performance of the operation will drive a new kind of collaborative performance improvements.

The DPMs should be aggrandized up through the asset sets to the entire plant. These aggrandized measures should become the primary performance measures for all levels in the organization charged with operational and business performance improvement. This should include engineering, supervision, management, planners, schedulers, and other operationally responsible personnel. Developing and aggrandizing the DPMs in this manner will ensure operational alignment throughout the organization, which will lead to higher levels of collaboration across all teams.

The third step is developing an appropriate incentive program to reward both operations and maintenance based on financial improvements in asset value. Implementing incentives for frontline personnel is anticultural for many industrial organizations, but incentives tend to help establish the fact that management is serious about performance. The incentives do not necessarily need to be financial. They could be as simple as creating a positive competitive environment that recognizes combined operations and maintenance teams. Utilizing performance measures and incentives in this manner helps drive collaboration between operations and maintenance. Engineering will tend to help overcome technical barriers to performance improvements if there are measures utilizing the same basic performance measures or aggrandized measures.

The fourth step is to systematize the asset performance management approach across each plant and then across the entire industrial enterprise by sharing and driving best practices from plant to plant. At each step in this progression, the incremental financial value generated by taking an asset performance approach should easily cover the cost of implementation. Since the DPMs would have been installed and the financial performance improvement made visible, it will be easy to discern the incremental financial improvements. This visibility should encourage management to continue investing.

Once the asset performance management system and culture have been established, the company can move to expand on asset performance management by moving to a control performance, people performance, or safety performance focus depending on the cultural biases of the company.

HUMAN PERFORMANCE EXCELLENCE

A number of industrial companies have developed an empowerment culture based on the execution of some form of continuous improvement (CI), such as total quality management (TQM), 6σ, or lean manufacturing. For these organizations, typically, the best path toward controlling the real-time enterprise would be to build on the empowerment culture by driving human performance excellence.

Since, as has been previously mentioned, all four paths to controlling the real-time enterprise overlap to a great extent, many of the basic concepts associated with human performance excellence have already been presented. A human performance excellence approach takes traditional CI programs to a higher level by extending the CI focus beyond team-based problem solving to individual problem solving in day-to-day operations. The initial focus of people performance excellence is on the actions and activities of frontline plant personnel who already make hundreds of decisions each day, with each decision having some impact on profitability. Although the profit impact of each decision may be fairly small, since they make so many decisions, the overall impact can be very significant to the business.

The objective of a human performance excellence program is transitioning the perspective of management and professional staff with respect to frontline personnel from "laborers" to "performance managers." Ever since the dawn of the Industrial Revolution, management and engineering perceived and treated frontline personnel as uneducated and unskilled laborers, at first, because they were. As the laborers' level of education, skill, and experience continually increased over the last 50 years, the perspectives of management and engineering do not seem to have changed at all. Today, the operators and maintenance people in many industrial organizations are more educated, experienced, and capable than ever. It is time for industrial companies to truly capitalize on this untapped resource. As Peter Drucker of Stanford University once pointed out, "Most discussions of decision making assume that only senior executives make decisions and that only senior executives' decisions matter, this is a dangerous mistake."

A good starting point in the transition toward a human performance excellence environment is with a general evaluation of the structure of the common CI programs being instituted in many industrial environments. These programs have evolved from work pioneered by Walter Shewhart in the 1920s and two of his disciples, W. Edwards Deming and Joseph Juran. All three of these innovators made significant contributions in the fields of quality improvement and CI. It is certainly beyond the scope of this book to review all aspects and advancements in the quality movement over the last century, but the basic concepts were really quite simple and straightforward applications of control theory. In order to improve the performance of any manufacturing operation, the performance had to be measured, the reasons for less-than-expected performance identified, and the root causes for underperformance eliminated—a simple application of control theory. As the root causes for underperformance were eliminated, the performance would naturally improve.

Most CI programs instituted in industrial companies have realized significant benefit, but fall short of the potential envisioned by the CI pioneers. Deming and Juran both encouraged CI programs based on strong management commitment and the use of CI teams to identify and alleviate the root causes of underperformance of industrial operations. However, they also encouraged a third component, quality improvement in daily work activities (Figure 7.8).

Today, many CI programs have had great success focusing on the first two aspects of CI; management commitment and CI team problem solving, but few have successfully transitioned the programs to CI in daily work. Perhaps, the reasons for this are both cultural and technical. The main cultural reason appears to be the distrust of frontline personnel by management and engineering. For CI in daily work to be effective, it would need to be implemented in a bottom-up manner. This means that frontline personnel would need to be the first to be empowered to make performance impacting decisions.

A clear analogy can be made between the board operators in, for example, a refinery or chemical plant, with airline pilots. Pilot (operator) error can cause disasters, but the flip side is also true. Operators and pilots have

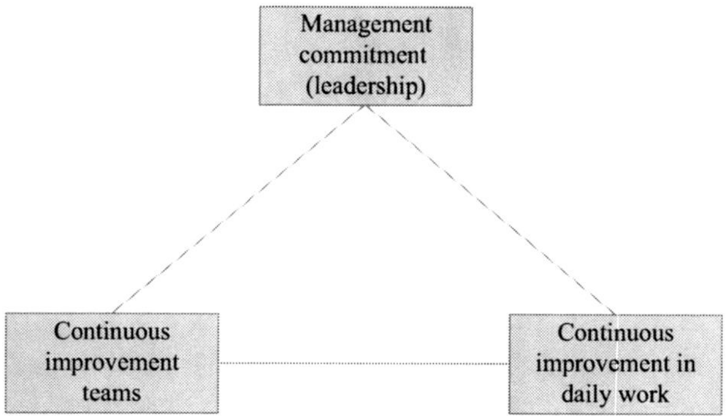

Figure 7.8. Continuous improvement requirements. There are three generally accepted requirements for an effective continuous improvement program: management commitment, continuous improvement in teams, and continuous improvement in individuals' daily work. Real-time business performance measurement and empowerment enable both team and individuals continuous improvement.

all the responsibility of driving your multimillion dollar asset, but operators do not have the authority pilots have to make changes based on real-time information.

The technical reason is that it has been very difficult to get the real-time performance measurement feedback necessary to underpin a CI in daily work approach. Empowering people requires that the information be provided to them in a timely manner that enables them to discern the performance impact of their actions. This would necessitate a real-time decision support infrastructure, which has been difficult to develop. The previously discussed DPM system with dashboard and scorecard empowerment provides exactly the real-time decision support infrastructure required to underpin CI in daily work.

One of the basic tools utilized in modern CI programs is the Pareto chart (Figure 7.9). Pareto charts are used to identify potential root causes of underperformance in a prioritized manner. The causes to the left of the chart are those that contribute most to the underperformance while those to the right have less impact. CI teams are established to execute a problem-solving exercise for the highest-impact root causes with the intention of eliminating the cause and thereby improving the performance of the operation. CI teams tend to follow very structured and well-proven problem-solving processes that may involve months of work for each of the root causes. Typically, CI teams are assigned to address the root causes from left to right along the Pareto charts, and at some point, the cost of the CI team problem solving is deemed to be too much for the expected benefit, and the CI activity may move to another

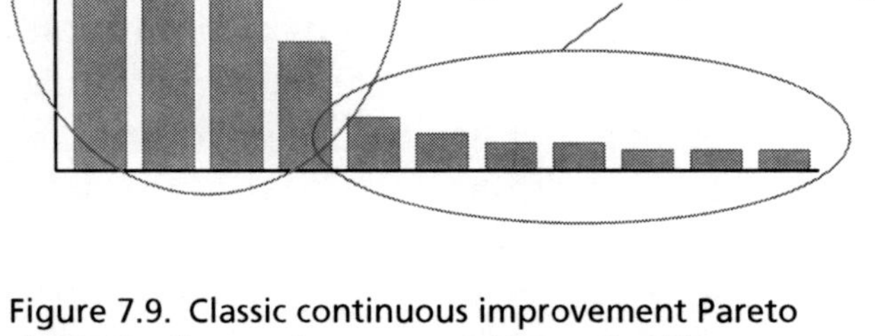

Figure 7.9. Classic continuous improvement Pareto chart. Continuous improvement opportunities are typically displayed on Pareto charts based on their value impact. The top opportunities tend to be addressed by CI teams but the real-time business measurement and empowerment enable smaller opportunities to be addressed through individual daily work.

major category of improvement. This leaves a number of minor root causes in the Pareto chart unaddressed. Although the potential value of each of these root causes may be small, added together, they may have a significant performance impact on the operation.

Extending a traditional CI program with a human performance excellence program based on CI in daily work serves to address two shortcomings of traditional CI programs. First is addressing the lower impact root causes in the operation that CI teams tend to miss. Second is actually sustaining the performance improvements gained through traditional CI team problem-solving exercises. Unfortunately, although many CI teams produce excellent results, experience has shown that the initial improvements can decline over time. Part of the reason for this decline is that once the team is on to the next project no one is focused on sustaining the improvement gains previously realized. An effective real-time decision support system, such as that proposed in a DPM-based approach, provides ongoing visibility to the performance measures. If the performance begins to decline, it will be immediately visible to the operation and the appropriate response can be taken. Visibility tends to lead to sustainability.

A fully envisioned CI program includes all three fundamental components: management commitment, CI team problem solving, and CI in daily work can be an extremely powerful engine for continuous and sustained performance improvements.

Industrial companies with a strong CI culture can make the transition to a human performance excellence approach by following a simple four-step process that will add an effective CI in the daily work component to their existing CI processes.

The first step is to implement a real-time measurement system for the performance measures of the operation as defined by the senior management. This is the essence of the DPM system previously discussed.

The second step is to use the prioritized DPMs to empower the frontline personnel with the exact real-time feedback information they require to understand the impact of their daily decisions on the performance of the operation within their domain of responsibility. The performance measures should be prioritized for the frontline personnel so that they can clearly discern which measures are most important. This can simply be accomplished by ordering the measures on the dashboard or scorecard displays by priority from left to right. This is providing the right information to the right people at the right time for them to be able to use it effectively.

It is also essential that engineering and management review their position on the decision rights of the frontline personnel. In many operations, the engineers work to "protect the process from the operators" by limiting the range of potential decisions the operators can make. In some cases, these limits are necessary for safety and environmental impact reasons, but this may not always be the case. If the operational personnel have the right decision support environment but are prevented from making the necessary decisions it can severely limit the impact of the system. This is not a simple issue. Decision rights must be very carefully considered to both maintain a safe environment and drive performance improvements.

The third step in the process is simply to train the operations personnel on what the real-time decision support environment is all about, their decision rights, and how to use the system as they perform their daily functions. The actual daily activities they perform within a people performance excellence environment may not change at all from what they traditionally have done, but the actual actions they take will. They will understand the impact of their actions and will learn how to perform their traditional activities in a manner that improves the performance of the operation.

The fourth step in the transition process is to determine how to reward the performance of the operational personnel that is having a positive impact on the performance of the operation. Many industrial organizations resist incentive programs right to the frontlines but incenting good behavior and good results can create a very positive work environment that supports a CI program. The incentives do not need to be monetary.[1] Additional benefits and simple recognitions can be extremely effective. If management wants to truly drive a performance culture in their organization they should be willing to consider incentives. Employees see incentive programs as a visible sign of management commitment to performance and to their people.

Once the frontline personnel are empowered, the human performance excellence process should systematically proceed up the levels of organization. This can and should be done by aggrandizing the DPMs up through the organizational nodes in a reverse manner from the initial Vollmann Decomposition (Figure 7.10). Following this recomposition process ensures that the organization is well aligned. This will encourage a much more supportive management structure, teaming, and collaboration. As the real-time decision support system progresses up through the organizations, performance improvements should continually increase. Peter Drucker encouraged this approach when he commented that, "Making good decisions is a crucial skill at every level." This process will ensure that the organization is making better decisions at each level and that management is aligned right down to the frontline performance managers of the organization.

---

[1] There is some discussion about the ancillary requirements for making other incentives greater than money in employee value. It appears that if the working environment is good, other, non-monetary incentives are valued greater than just money. If the working environment is poor, most employees would prefer more money.

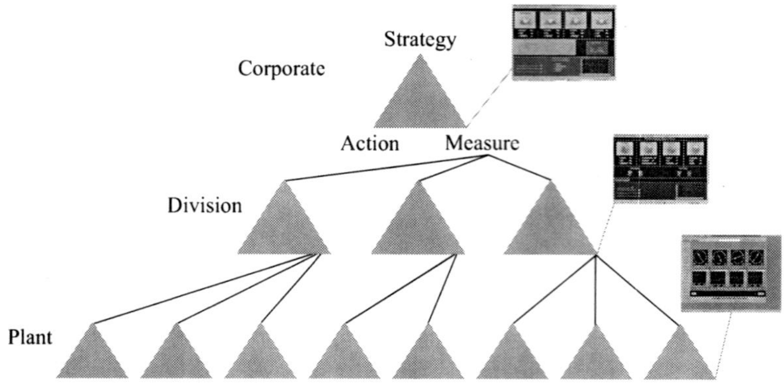

Figure 7.10. Multilevel people performance excellence. Executing a top-down Vollmann Decomposition analysis and a bottom-up recomposition-based implementation drives real-time empowerment and performance excellence at all levels across the entire operation.

Industry is currently experiencing a talent crisis. Many industrial workers—both frontline and professional—are Baby Boomers who are getting ready to retire. The downturns in the industrial markets of the past 20 years led to both downsizing and hiring freezes. This means that there is a talent gap across industry with a number of experienced employees older than 55 years, a number of less experienced employees younger than 35 years, and a few employees between 35 and 50 years. This will cause increasing challenges to industrial companies over the next 5–10 years. To exacerbate this situation, fewer and fewer people are finding industrial jobs to be as interesting as high technology and business jobs. Moving to a true human performance excellence approach will help to resolve these talent crisis issues by helping in creating more attractive work experiences, bringing new talent on in a shorter time, increasing morale, and improving the value from the employee base.

Experience has demonstrated that the time to get new employees to appropriate levels of effectiveness can be shortened significantly by coupling a human performance excellence program with state-of-the-art training utilizing training simulators—similar to those used for pilot training.[2] This can be supplemented by utilizing virtual reality software that enables staff to be continually trained on how to perform their activities more effectively. Once they are up to speed, utilizing the real-time decision support environment will actually cause them to perform at higher levels than the talent they are replacing.

[2] Data indicate that operators trained conventionally take a minimum of two years to become capable of running a control system board while simulator-trained operators can be turned loose in around six months.

It is amazing what people can accomplish given the change and the right tools. People tend to like to contribute to success and know that they have had a positive impact in the jobs they perform. True empowerment enables them to succeed and increases enthusiasm and morale. This can become infectious leading to highly attractive work environments. The new technology-based environments and appropriate incentives that come from a people performance excellence approach should help industrial companies to attract and retain top talent into their organizations and help stave off the talent crisis facing industry.

SAFETY AND ENVIRONMENTAL PERFORMANCE EXCELLENCE

With increasing regulatory pressure and a number of highly publicized safety events impacting industry, a number of industrial companies have increased their focus on environmental stewardship and safety. As a result, these companies have a strong safety culture. Since, as has been previously discussed, safety and environmental concerns together tend to provide constraints on profitability, the word "safety" as used in this section will refer to safety of people, plant, and environment. Because of the constrained nature of safety, building on a safety culture to transition to RTEC may be a less direct path than control excellence, asset excellence, and people excellence, but safety excellence can still be effectively leveraged.

A number of safety-related approaches have been implemented that serve as a necessary and sound basis for safety excellence. In the heavy process industries, specific safety shutdown systems and safety instrumented systems provide necessary process safety safeguards that must carefully be considered and implemented where appropriate. In automated batch manufacturing environments, the exception logic associated with batch processes must be carefully implemented to provide appropriate safety protection. Other safeguards, such as pressure relief valves and fail-safe control valves, which have been utilized for many decades, must be correctly designed and implemented into industrial processes. Operator decision rights ranges need to be effectively built into control strategies for all processes to make sure that operators do not take any unsafe actions. Similar environmental measurements and control limits must also be built into industrial processes. All of these traditional approaches to safety have been developed over the years as a baseline for the safe operation of industrial processes. For the remainder of the safety excellence discussion these safeguards will be assumed to be effectively deployed.

Safety is critically important in any industrial company. However, safety is typically not any profit-based company's primary objective. If it was it could easily be achieved by turning off the plant. Maximizing the production value of the operation is typically the main objective of most industrial operation. Minimizing the variable costs of the operation is typically a contributing objective. The components of variable cost that tend to fluctuate in real time are the cost of energy and the cost of raw material. Safety and environmental concerns provide constraints on profitability (refer Figure 7.1). Although it is not a good public relations strategy to admit this, it is useful to the discussion to truly understand the relationship between profitability and safety.[3]

This is not to say that industrial companies do not take safety seriously—they certainly do. I have never met an industrial executive who would not work diligently to ensure the safety of their employees, the plant equipment, and the community at large. However, many industrial processes are inherently unsafe. Just to manufacture some products that are required by society safety risks need to be taken. Unfortunately, this is a fact of industry. The total elimination of safety risk is not currently feasible for many industrial concerns. Therefore, the key to safety excellence is the effective balancing of safety risk and profitability in a manner that minimizes the probability of an unsafe event while enabling profitable production.

The real-time profitability model of Figure 7.1 presents a simplified perspective of the relationship between safety and profitability. This model is really oversimplified on a number of dimensions. First, safety risk does not present the only constraint to industrial profitability. Other constraints, such as equipment limitations, operator perspectives, engineering design, and maintained equipment state, also play a role. However, in most high-risk production processes, the constraints associated with safety and environmental integrity tend to be encountered before other constraints and thereby provide the most pressing constraints on profitability. For those processes in which other constraints are encountered before safety risk constraints, high safety risk conditions should not be encountered. Any industrial plant moving down a safety excellence path should focus on the high safety risk processes. The following discussions will assume that the processes under consideration have safety risk as the primary operational constraint on profitability.

Traditionally, industrial operations have appropriately taken a very conservative approach to the safety of their operations. Since the consequences of an unsafe event can be catastrophic, the tactic has typically been to operate the plant in a zone that will not exceed the safety envelope of the operation. The safety envelop is a major constraint associated with the operation that, if exceeds, will almost always lead to an unsafe event (Figure 7.11). The safety

[3] However, the "after crisis" costs of unsafe operations are never considered in the relationship between profitability and safety. BP's huge cost burden after Deepwater Horizon is the poster child for this. If BP's safety systems had worked, all those billions of dollars would be enhancing profit instead of acting as a financial anchor.

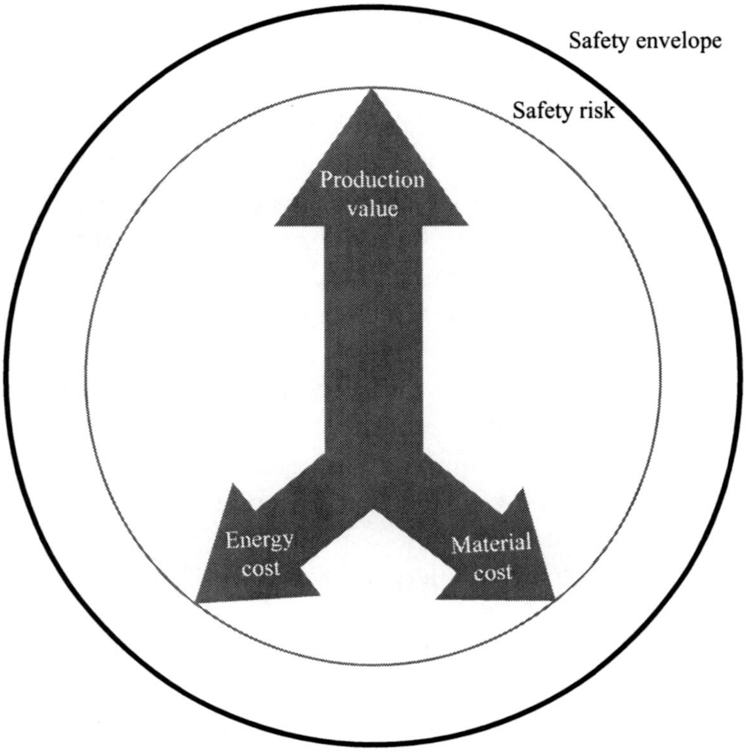

Figure 7.11. Safety risk versus safety envelope. The safety envelope of industrial operations represents the operating conditions that should not be violated without high potential for a safety incident. Most operations run well within the safety envelope in a domain at which the safety risk is acceptable.

envelope in an industrial operation can shift over time with equipment wear and mechanical stress, but it tends to be fairly fixed and stable. Since the safety envelope is typically determined by engineering analysis and is not directly measured, the primary goal of production is to operate the plant in a manner that will never lead to a violation of the envelope. Therefore, the plant is operated some distance from the envelope based on the feelings of the operations, maintenance, and engineering staffs based on reasonable plant operations while minimizing the risk of violation of the safety envelope. Therefore, there is a second operating envelope in most high-risk process operations called the safety risk envelope. This envelope is set well inside the safety envelope typically based on operating experience and engineering analysis.

For many industrial processes, the safety risk changes with different operating conditions or phases of operation. As an extreme example, during plant start-up, the safety risk tends to be greater than it is during normal operations. This is evident when developing the exception logic in automated batch processing applications. Normal batch processing is done in phases of operation and the exception logic is typically tied to the phase. This is because some phases are at higher risk than others and therefore require more exceptional handling. For example, in a chemical process involving an exothermic reaction, there may be very little risk during the production phase involving charging the reactor with the base chemicals needed, but during the catalyst charge phase, there may be a very high risk of an unsafe event. Therefore, the exception logic for the catalyst charge phase may be much more intensive than that for the chemical charge phase. In continuous processes, similar safety risk fluctuations may occur.

Since safety risk is not directly measured in most industrial processes, the safety risk envelope is typically established according to the worst safety risk potential the process may encounter. This makes very sound sense since that safety risk envelope will be the one that generally enables the operation to produce a product effectively while avoiding safety incidents efficaciously. Without direct measurement of safety risk this is the best compromise that can be made.

There may be an opportunity to safely drive additional profitability if the safety risk envelope could be directly measured for a production process. Over different operating conditions or different operating phases the actual safety risk may reduce, which would loosen the safety risk envelope (Figure 7.12). If this risk reduction was measured and made available to the operations staff, they may be able to capitalize on it by pushing the process, extending the real-time profitability vectors to increase the profitability of the operation. If the safety risk were to increase again, the operations team could back down on the process to bring it within a reasonable safety risk range. The basic result would be increased profitability from the processing without increasing the safety risk. Although the amount of profit increase may be fairly small as compared to the overall production value of the operation, a number of small increases can be the difference between an industrial firm being market competitive or not.

The importance of directly tying safety to profitability from the perspective of creating a safety culture cannot be overstated. From a business perspective, safety is an annoyance—as is any constraint on profitability. However, business executives understand that they need to maintain safe operations both as altruistic leaders and because a safe environment is often a prerequisite for a license to operate. To this point, the primary organization strategy to ensure safety has been to create an Environment, Health, & Safety (EH&S) department within their operations to oversee safety. EH&S teams are typically not viewed as part of the fundamental business organization, rather, they are often seen as an adjunct to the organization. This is not to say that they do not have significant authority, rather that they are not mainstream in most industrial companies. It could be of great benefit to industrial companies and to the safety

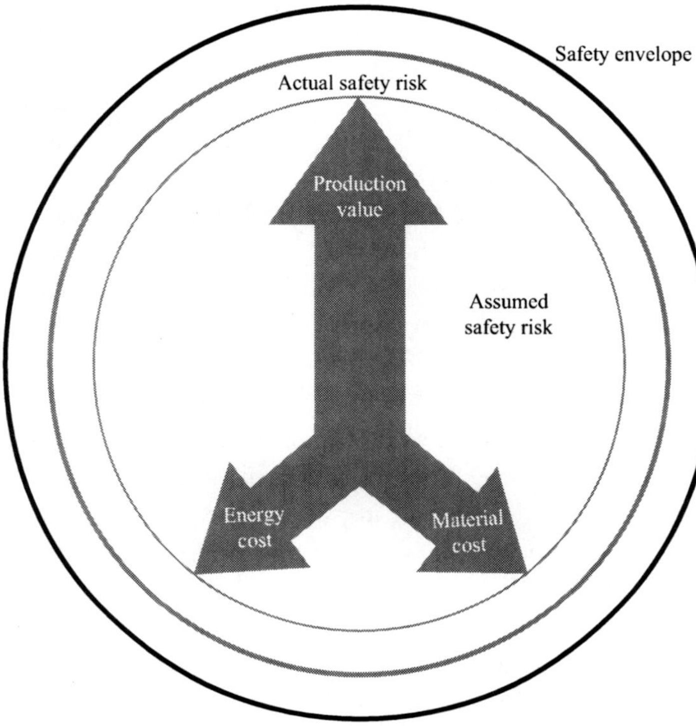

Figure 7.12. Safety envelope. Since the actual safety risk is seldom measured, the plant operates at an assumed safety risk, which is often more constraining on the operational profitability than is necessary. Measuring the actual safety risk in real time may enable profitability improvements while maintaining safe operations.

culture of these companies if the EH&S function was considered a mainstream business function. This can be accomplished by clearly developing the direct relationship between safety and profitability. Instead of safety being an insurance policy for the company, it can be viewed as a profit-impacting function. In many organizations, this can lead to much tighter safety integration.

Enabling this requires the safety risk of the operation to be measured in real time. This is a huge challenge that is still being developed, but some initial progress has been realized. Safety risk can be looked at from two different perspectives. Governmental organizations such as Occupational Safety and Health Administration (OSHA) develop standards for maintaining the safety of industrial operations based on regular and rigorous inspections. Experience has shown that industrial companies that are diligent in following the OSHA requirements very carefully have lower safety risk than companies that do not. For purposes of this analysis, this component of safety risk will be referred to as operational risk. Companies such as ACM Facility Safety™ in Calgary,

Alberta, Canada, have advanced the state-of-the-art in measuring operational safety risk to a great degree. Although the measurement of operational safety risk is not of high enough resolution to give it a specific numeric value yet, it can typically be set on a high, medium, and low scale based on the diligence of the organization with respect to the audits.

As good and effective as the inspections and audits are they may not catch all safety issues in an operation. There may be same safety events that occur in industrial operations that are invisible to inspections. These events are often based on an unexpected set of conditions that are met in the same time frame. If these conditions could be identified and recognized prior to the safety incident, the operation may be able to take action to avoid the incident. This aspect of safety risk is referred to as conditional safety risk. Conditional safety risk can be defined as the existence of a condition that may lead to a safety incident multiplied by the expected impact of the incident. The major challenge with conditional safety risk is identifying potential conditions and their lead indicators. One approach that may help in this involves mining the huge quantities of historical data on manufacturing processes that have been collected through process historian software over the last three decades. Many industrial companies need to keep these data as evidence of their production, but they do not tend to use the data to identify potential safety incidences and their lead indicators. There is huge potential in mining these data for safety incident identification purposes. If industrial companies were to identify past safety incidences and analyze the historical process data for the periods leading up to those incidences, they may be able to identify leading patterns and relationships that were not obvious to the operators. By identifying these patterns for past events they could set up pattern recognition software that scans actual operating conditions to determine if any similar patterns are arising in the operation. When a potential threat is recognized the software can notify the operations staff that the conditional safety risk is elevated and inform them of what the potential incident may be. In this way, the operations staff may be able to take actions to prevent the occurrence of an impending event.

As with the operations safety risk, conditional safety risk may not be able to be precisely measured from a numeric perspective. The risk levels may need to be limited in precision to low, medium, and high ranges, but conditional risk certainly will provide additional safety risk information to the operations team that may help them to manage additional profitability in a safe manner. One of the shortcomings of conditional safety risk measurement is that there may be potentially unsafe conditions that have never previously occurred at a plant. Therefore, the historical data analysis for that plant will be of no value in detecting the lead indicators of those potential incidences. Since many industrial plants have huge volumes of historical data, this gap may be able to be closed to some degree by analyzing incidences that have occurred at other

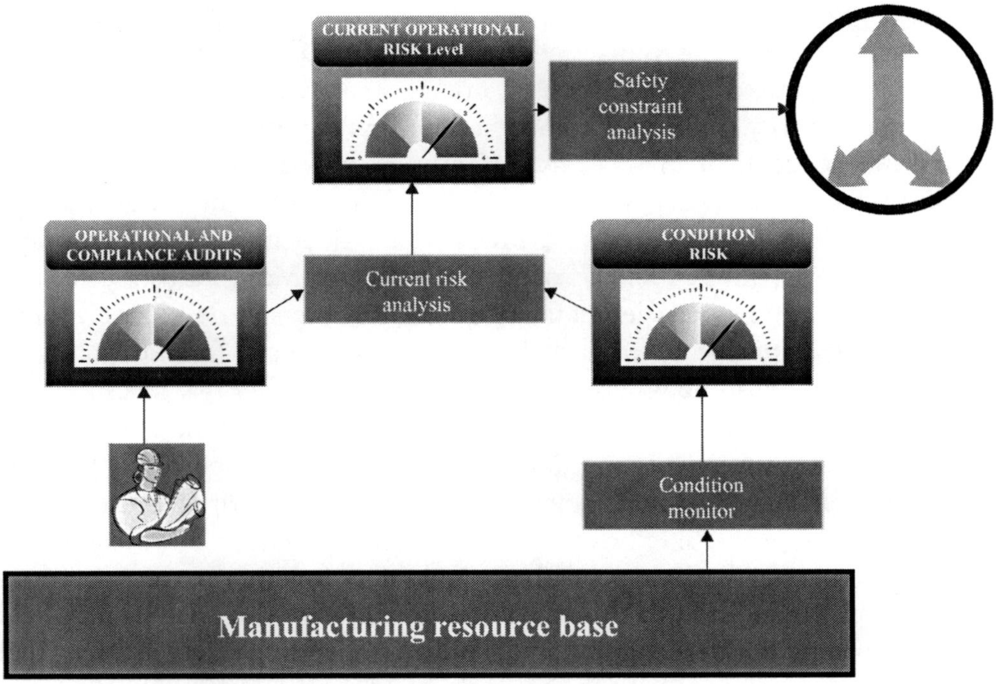

Figure 7.13. Safety risk measurement. A perfectly accurate real-time measurement of safety risk approach may not exist for most operations, but a conservatively derived composite safety risk measurement based on the highest risk of operational and compliance and conditional safety risks may help to move the science—and profitability—forward in a reasonable manner.

similar industrial plants and seeing if the lead indicators of those incidences may be able to be extrapolated to other plants. Certainly, this is not fool proof, but it may help industrial companies to avoid a large number of unsafe events.[4]

Communicating the safety risk of the operation to operators should be done in a very conservative manner as is typically the case when dealing with safety. Providing the operator with separate dashboard elements for operational and conditional safety risk may only serve to confuse the situation. When dealing with safety, confusion is not good. One suggestion would be to combine the operational and safety risk scales into an overall safety risk indicator (Figure 7.13). This can simply be accomplished by taking the higher of the two risk measures and using it as the overall risk indicator. This is totally consistent with industries' conservative approach to safety and it certainly passes the common-sense test.

[4] Another lesson from the BP Texas City incident is that they were unaware of the more than 10 near-miss incidents that occurred before the actual accident. One of the requirements for real-time assessment of safety is improved data about as-built systems and the importance of improved information transfer.

More research and experimentation are required to support the move to safety excellence, but the initial work that has been done is very promising. As was previously stated, "safety" as used in this discussion also applied to environmental integrity. Industrial companies are striving for safety and environmental excellence for many reasons. Safety and environmental incidences can have huge negative impact on companies in the form of injury, death, equipment and facility damage, reduced production value, business interruption, lost market opportunities, and certainly bad publicity. Many industrial companies clearly recognize the value of safety excellence and have developed a safety culture. This safety culture may provide an ideal launch pad for a drive toward controlling the real-time enterprise. Using real-time safety risk measurement indicators, the primary constraints on profitability in many industrial operations can be measured to a good degree. If the safety risk is currently low, the operations staff may find that they can drive the process a little harder to get some additional profitability. As the safety risk increases, they may need to cut back. This will effectively tie profitability and safety together in the operations team's mind. In addition to this, the conditional safety risk analysis may help industrial companies to recognize and avoid an impending safety incident that will even have more impact on the profitability of the operation. Although the move to safety performance excellence may not be as developed as the other three, the benefits can be just as significant.

SUMMARY

The business of industry is undergoing a huge transformation from high stability over extended time frames to real-time variability. While the efficiency of industrial plants may be well controlled, more and more executives are frustrated that the profitability of their operations appears to be—and is—out of control. This change in market dynamics is presenting daunting challenges to industrial companies. They have often tried to rise to these challenges by trying to get more decision support information from their Enterprise Resource Planning (ERP) systems. Unfortunately, the required real-time decision support information is just not there. The speed of business is faster than the financial reporting system's ability to keep up.

This situation can, on initial inspection, appear to present a conundrum to industrial management. However, the solution to this situation is to identify the actual type of problem that industry is facing and to solve the problem accordingly. The problem is a real-time control problem. Nevertheless, it is not based on just controlling the efficiency of the industrial plants. It is based on the need to control the profitability of industrial businesses. Accomplishing this will require talent that understands business, real-time computing, and feedback control. Fortunately, that talent exists in most industrial organizations. The plant engineers understand real-time computing and feedback control while business management and information technology talent understand the business implications. By combining these talent sets, the solution to this daunting problem is at hand.

Applying real-time control theory to both the plant operations and industrial businesses is referred to as controlling the real-time enterprise. Transitioning to RTEC is not easy, but companies can start within their cultural biases and move from there. This will enable them to move forward in a more comfortable and effective manner. The path is not easy, but the results will be well worth the effort.

Bibliography

A Special Report on Factory Automation. "Linking the Pieces of CIM," *IndustryWeek*. Cleveland, OH: Penton Publishing, March 23, 1987.

A Special Report on Factory Automation. "The U.S. and Quality: A New Culture," *IndustryWeek*. Cleveland, OH: Penton Publishing, April 17, 1989.

A State-of-the-Art Report. "Factories of the Future," *IndustryWeek*. Cleveland, OH: Penton Publishing, March 21, 1988.

Abel, J.; and P. Martin. "Automation, Business, and Operating Advances Align into New Paradigm for Economic Performance Improvement," *Pharmaceutical Engineering*. Tampa, FL: ISPE, January/February 2000.

Allen, L.; and D. Stovicek. "CIM Goes to School," *Automation*. Cleveland, OH: Penton Publishing, May 1990.

Babb, M. "Process Control Systems in the 1990s Will Enter a 'New Age'," *Control Engineering*, Chicago, IL, . Cahners Publishing, January 1990.

Badavas, Dr. P.C.; and Dr. A.D. Epperly. *Statistical Process Control Integrated with Distributed Control Systems*, 1988 NPRA Computer Conference, Pittsburgh, PA, October 1988.

Baer, T. "The Software Side of Quality," *Managing Automation*. New York, NY: Thomas Publishing Company, May 1990.

Balch, B.; and T. Miller. "Statistical Process Control in Food Processing," *Intech*, July 1990.

Beaverstock, M.; and L. Byzyna. "Measuring Performance Dynamically," *Control Engineering*. Chicago, IL, Cahners Publishing, June 1990, Volume 11.

Benassi, F. "The Long Road to Quality," *Managing Automation*. New York, NY: Thomas Publishing Company, May 1990.

Box, G.E.P.; W.G. Hunter; and J.S. Hunter. *Statistics for Experimenters: An Introduction to Design, Data Analysis, and Model Building*. New York, NY: John Wiley & Sons, 1978.

Brimson, J.A. "How Advanced Manufacturing Technologies are Reshaping Cost Management," *Management Accounting*, March 1986.

Brown, M.E.; K. Parker; and Jr. C.P. Senyard. "Computer Integrated Manufacturing in the Chemical Process Industries," *Chemical Processing*, October 1989.

Bunnell, J.; and R.L. McAllister. "Continuous Improvement Through Plantwide Integration," *Control Engineering* Chicago, IL. Cahners Publishing, June 1990, Volume II.

Bylinsky, G. "Challengers Move in on ERP," *Fortune*. New York, NY: Time Life Inc., November 22, 1999.

Clark, K.B.; and R.H. Hayes. "Why Some Factories are More Productive Than Others," *Harvard Business Review*. Cambridge, MA: September–October 1986.

Cobble, C.; and S. Player. *Cornerstones of Decision Making: Profiles of Enterprise ABM*. Greensboro, NC, Oakhill Press, 1999.

Conneally, D.; D. Lovett; and P.G. Martin. "An Econometric Approach to Justifying Advanced Process Control Projects," *Control Solutions*, March 2001.

Cooper, R.; and R.S. Kaplan. "Measure Costs Right: Make the Right Decisions," *Harvard Business Review*, September–October 1988.

Cooper, R.; and P.G. Martin. "Real-Time Cost Accounting: From the Factory Floor to the Bottom Line," *Baan Webcast*, October 2001.

Cooper, N.; and P.G. Martin. "The Road to Asset Performance Management," *Plant Services Webcast*, February 2007.

Corbett, T. *Throughput Accounting: TOC's Management Accounting System*. Great Barrington, MA: The North River Press Publishing Corporation, 1998.

Cox, B.J. *Object Oriented Programming, An Evolutionary Approach*. Reading, MA: Addison-Wesley Publishing Company, 1986.

Crosby, P.B. *Quality is Free: The Art of Making Quality Certain*. New York, NY: McGraw-Hill Book Company, 1979.

Deming, W.E. *Out of the Crisis*. Cambridge, MA: MIT, Center for Advanced Engineering Study, 1986.

Drucker, P.F. *Innovation and Entrepreneurship: Practices and Principles*. New York, NY: Harper and Row, 1985.

Drucker, P.F. *The Practice of Management*. New York, NY: Harper & Row, 1954.

Drucker, P.F. "The Emerging Theory of Manufacturing," *Harvard Business Review*, May–June 1990.

Elwart, S.P.; and P.G. Martin. "New Software Structures Extend Control Capabilities," *Control Engineering*, Cahners Publishing, June 1990, Volume II.

Friedman, P.G. *Economics of Control Improvement*, ISA, Research Triangle Park, NC, 1995.

Ghosh, A.; and H.P. Rosenof. *Batch Process Automation: Theory and Practice*. New York, NY: Van Nostrand Reinhold Company, Inc., 1987.

Goldratt, E.M. *What Is This Thing Called Theory of Constraints and How Should It be Implemented?* Great Barrington, MA: The North River Publishing Corporation, 1990.

Goldratt, E.M. *Necessary But Not Sufficient: A /Theory of Constraints Business Novel*. Great Barrington, MA: The North River Publishing Corporation, 2000.

Gould, L. "The CIM Task Force: Purchasing CIM is also a Distributed, Hierarchical, Pursuit," *Managing Automation*, December 1990.

Grant, E.L.; and R.S. Leavenworth. *Statistical Quality Control*. 5th ed. New York, NY: McGraw-Hill Book Company, 1980.

Harbor Research. "Performance Measurement: Impact on Competitive Performance," *Outlook, Industrial Competitiveness* Vol. 6, no. 4. Boston, MA: Harbor Research Corp., 1990.

Harrold, D. "Enterprise Integration Requires Understanding of the Plant Floor," *Control Engineering*. Chicago, IL, . Cahners Publishing, February 2000.

Harry, M.J., Ph.D. *The Vision of Six Sigma: A Roadmap for Breakthrough*, Tempe. AZ: Six Sigma Academy, Inc., 1994.

Hays, W.L. *Statistics for the Social Sciences*. New York, NY: Holt, Rinehart and Winston, Inc., 1973.

Henderson, B.A.; and J.L. Larco. *Lean Transformation*, Richmond, VA: The Oaklea Press, 1999.

Hidden, A.E.; and E. Lowe. *Computer Control in the Process Industries*. London, UK: Peter Peregrinus Ltd., 1971.

IBC Technical Services. *Computer Integrated Manufacture in the Process Industries-Conference*. London, UK: IBC Technical Services Ltd., 1990.

Jasany, L.C. "Knowledge (and Power) to the People," *Automation*. Cleveland, OH: Penton Publishing, July 1990.

Jones, D.T.; D. Roos; and J.P. Womack. *The Machine That Changed the World: The Story of Lean Production*. New York, NY: Harper Perenial, A Division of Harper-Collins Publishing, 1991.

Juran, J.M. *Managerial Breakthrough: A New Concept of the Manager's Job*. New York, NY: McGraw-Hill Book Company, 1964.

Juran, Dr. J. *Juran on Leadership for Quality—An Executive Handbook.* New York, NY: The Free Press, 1989.

Kaplan, R.S. "Accounting Lag: The Obsolescence of Cost Accounting Systems," *Harvard Business School 75th Anniversary Colloquium on Productivity and Technology.* Cambridge, MA: Harvard Business School, March 1984.

Kaplan, R.S. "Yesterday's Accounting Undermines Production," *Harvard Business Review,* July–August 1984.

Kaplan, R.S. "One Cost System Isn't Enough," *Harvard Business Review.* Cambridge, MA: January–February 1988.

Kaplan, R.S.; and D.P. Norton, *The Strategy-Focused Organization: How Balanced Scorecard Companies Thrive in the New Business Environment.* Boston, MA: Harvard Business Press, 2001.

Kompass, E.J. "The Road to Plantwide Information," *Control Engineering.* Vol. 2. Chicago, IL Cahners Publishing, June 1990.

Lareau, A.F. "Bringing Information to Automation," *Control Engineering.* Vol. 2. Chicago, IL, Cahners Publishing, June 1990.

MacGregor, J.F. "On-Line Statistical Process Control," *Chemical Engineering Progress,* October 1988.

Martin, P.G. "A Leadership Dilemma," *InTech,* February 2008.

Martin, P.G. "A New Blueprint for Automation," *Chemical Plants and Processing,* September 1989.

Martin, P.G. "A Software Structure for Open Industrial Systems," *Proceedings of the Fourteenth Annual Advanced Control Conference,* Purdue University and Control Engineering, September 1989.

Martin, P.G. "Achieving Process Manufacturing Excellence," *IndustryWeek Webcast,* November 2007.

Martin, P.G. "Automation Outlook," *Start,* February 2002.

Martin, P.G. "Automation Payback," *Industrial Computing,* Research Triangle Park, NC: ISA, August 1999.

Martin, P.G. "Bottom Line Automation," *ISA,* October 2001.

Martin, P.G.; and M. Turk. "Bottom to Top Real Time Accounting and Performance Measures: From Plant Floor to ERP," *NPRA,* October 2008.

Martin, P.G. "Computer Control of Batch Processes," *Measurement and Control,* Issue 106, September 1984.

Martin, P.G. "Designing a Batch Control System," *I&CS.* Radnor, PA: Chilton Company, October 1987.

Martin, P.G. "Don't Fumble Your Tech Funding Opportunity,"_*Pharmaceutical Manufacturing*, January 2009.

Martin, P.G. "Driving Bottom Line Value through Enterprise Control," *Heartbeat*, December 2008.

Martin, P.G. *Dynamic Performance Management: The Path to World Class Manufacturing*, New York, NY: Van Nostrand Reinhold, 1993.

Martin, P.G. "Interoperability is Not the Solution," *InTech*, October 2005.

Martin, P.G. "Living the dream of plant asset management," *InTech*, October 2005.

Martin, P.G. "Manufacturing Intelligence through Enterprise Control," *Manufacturing Business Technology Webcast*, October 2007.

Martin, P.G. "Metrics that Matter," *Managing Automation Webcast*, June 2005.

Martin, P.G. "Offene Industriesysteme fur die Verfarhrenstechnik-Schritt in ein neues Automatisierungszeitalter," *Verfahrenstechnik, Nr.* 10, 1989.

Martin, P.G. "OIS: A Blueprint for Automation," *Control Engineering* Chicago, IL: Cahners Publishing, February 1989.

Martin, P.G. *Open Communications in the Process Industries*, Paper presented at Instrument Society of America, Houston Channel Section, April 1990.

Martin, P.G. "Open Industrial Systems," *Measurement and Control,* October 1989.

Martin, P.G. "Out of Many ... One: Pointing the Whole Organization in the same Direction," *UpTime*, January 2009.

Martin, P.G. "Real-Time Performance Measurement and Management: From the Plant Floor to the Bottom Line," *AIChE 2003 Spring National Meeting*, April 2003.

Martin, P.G. "Temperature, Pressure, Flow, Profit," *InTech*, February 2001.

Martin, P.G. "The Benefits and Value of Automation," *ISA*, October 2008.

Martin, P.G. "The Business ofEngineering," *Industrial Management*, January/February 2004.

Martin, P.G. "The Information Empowered Enterprise: Driving Operational Excellence," *Control Engineering*, January 2008.

Martin, P.G. "The Move to Open communications in the Process Industries," Paper presented at Instrument Society of America, Calgary Section, April 1990.

Martin, P.G. "The Move Toward Open Industrial Systems," *Proceedings of the Advanced Control Symposium,* Texas A&M University, February 1989.

Martin, P.G. "The Move to Performance-Based Automation," *InTech*, September 2004.

Martin, P.G. "Time for Leaders to Step Forward," *InTech*, September 2008.

Martin, P.G.; and V. Bapat. "The Pursuit of the Perfect Plant: Enabling Real Time Production Responsiveness through Enterprise Control," *Automation World Webcast*, October 2007.

Maskell, B.H. "Performance Measurement for World Class Manufacturing, Part I," *Manufacturing Systems*. Carol Stream, IL: Hitchcock Publishing Co., July 1989.

Maskell, B.H. "Performance Measurement for World Class Manufacturing,

Part II," *Manufacturing Systems*. Carol Stream, IL: Hitchcock Publishing Co., August 1989.

McLuhan, M. *The Mechanical Bride: Folklore of Industrial Man*. Boston, MA: Beacon Press, 1952.

Miller, J.G.; A.J. Nianni; and T.E. Vollman. "What Shall We Account For?" *Management Accounting*. January 1988.

Miller, J.G.; and T.E. Voliman. "The Hidden Factory," *Harvard Business Review*, September–October 1985.

Mitchell, J.S. *Physical Asset Management Handbook*. 4th ed. Houston, TX: Clarion Technical Publishers, 2006.

Murrill, P.W. *Fundamentals of Process Control Theory*, Research Triangle Park, NC: Instrument Society of America, 1981.

Passino, R. "Computers Spur Profitability at General Chemical," *Chemical Processing*, Chicago, IL: Putman Publishing, November 1990.

Pastor, E. "Networks: The Backbone of Integration," *Automation*, Cleveland, OH: Penton Publishing, May 1990.

Perrow, C. *Normal Accidents: Living with High-Risk Technologies*, Princeton, NJ: Princeton University Press, 1999.

Peters, T.J.; and Jr.R.H. Waterman. *In Search of Excellence: Lessons from America's Best-Run Companies*. New York, NY: Harper & Row, 1979.

Qualtec, Inc., *Team Leader Training Course, Participant Workbook*. Miami, FL: Florida Power & Light Co., 1987.

Qualtec, Inc., *Managing Quality Improvement, Participant Workbook*. Miami, FL: Florida Power & Light Co., 1988.

Rohan, T.M. "Factories of the Future," *IndustryWeek*, March 21, 1988.

Rossi, D.A. "Consider Integrated Solution for Total Quality," *Intech*, January 1990.

Shaw, W.T. *Computer Control of Batch Processes*. Cockeysville, MD: EMC Controls, Inc., 1982.

Shewart, W.A. *Economic Control of Quality of Manufactured Product.* New York, NY: Van Nostrand Company, Inc., 1931.

Shinskey, F.G. *Process-Control Systems.* New York, NY: McGraw-Hill Book Company, 1979.

The Foxboro Company. *Introduction to Process Control.* Foxboro, MA: 1986.

Trevathan, V.L. Editor. *A Guide to the Automation Body of Knowledge.* Research Triangle Park, NC: ISA, 2006.

Walsh, S. "Perspectives on the Age of the Smart Machine," *Managing Automation,* December 1989.

Williams, T.J. *A Reference Model for Computer Integrated Manufacturing (CIM): A Description from the Viewpoint of Industrial Automation.* Research Triangle Park, NC: Instrument Society of America, 1989.

Zuboff, S. *In the Age of the Smart Machine.* New York, NY: Basic Books, Inc., 1988.

Index

CPSIA information can be obtained at www.ICGtesting.com
Printed in the USA
BVOW03*1633090614

355586BV00002B/2/P

9 781606 503591